**ROUTLEDGE LIBRARY EDITIONS:
POLLUTION, CLIMATE AND CHANGE**

Volume 5

ATMOSPHERIC PROCESSES

ATMOSPHERIC PROCESSES

JAMES HANWELL

Routledge
Taylor & Francis Group

LONDON AND NEW YORK

First published in 1980 by George Allen & Unwin

This edition first published in 2020
by Routledge
2 Park Square, Milton Park, Abingdon, Oxon OX14 4RN

and by Routledge
52 Vanderbilt Avenue, New York, NY 10017

Routledge is an imprint of the Taylor & Francis Group, an informa business

British Library Cataloguing in Publication Data
A catalogue record for this book is available from the British Library

ISBN: 978-0-367-34494-8 (Set)
ISBN: 978-0-429-34741-2 (Set) (ebk)
ISBN: 978-0-367-36215-7 (Volume 5) (hbk)
ISBN: 978-0-367-36220-1 (Volume 5) (pbk)
ISBN: 978-0-429-34464-0 (Volume 5) (ebk)

Publisher's Note
The publisher has gone to great lengths to ensure the quality of this reprint but points out that some imperfections in the original copies may be apparent.

Disclaimer
The publisher has made every effort to trace copyright holders and would welcome correspondence from those they have been unable to trace.

ATMOSPHERIC PROCESSES

James D. Hanwell
The Blue School, Wells,
Somerset

London
GEORGE ALLEN & UNWIN
Boston   Sydney

GEORGE ALLEN & UNWIN
40 Museum Street, London WC1A 1LU

© J. D. Hanwell, 1980

British Library Cataloguing in Publication Data

Hanwell, James David
    Atmospheric processes. – (Processes in physical
geography; no. 3).
    1. Atmosphere
    I. Title II. Series
    551.5   QC861.2   79–40955

    ISBN 0–04–551032–6

Typeset in 10 on 11 point Times by Northampton Phototypesetters Ltd
and printed in Great Britain by
Hazell Watson & Viney Ltd, Aylesbury, Bucks

Preface

As we probe space further to explore neighbouring planets, the key question posed first always concerns the atmospheres that envelope them. Information about their composition and behaviour reveals the exciting possibility of life elsewhere in the Universe. Earth's atmosphere, for example, is vital in sustaining all terrestrial life, and its processes strongly influence all living and non-living features of the landscape and oceans. The obvious outcome of these processes is the great variety of weather experienced around the world. So, the harmony between processes operating on the surface and in the sky affects us all, whether we are geographers, farmers, builders, sportsmen, politicians or the members of countless other occupational groups.

In schools, the study of the atmosphere is interwoven into geography courses and studies of the natural or *physical* environment. A great deal of this has involved descriptive climatic statistics and the necessarily somewhat arbitrary rules for classifying different régimes in various ways. As the basis for learning about the atmosphere, the former especially does little to help either the student's understanding of the weather or the subject's scope in relating Man and his environment.

It is the writer's experience in teaching the subject to sixth formers for over twenty years that an approach through studying *processes* makes the atmosphere come alive, for there is always something to look at in the sky and to ponder upon. To the reader, then, this is not a book of data to be memorised, but of the drama in the air we breathe and our surroundings. To the expert, the leaning is towards physical meteorology and climatology with the props of the dynamic and synoptic approaches where they are helpful.

In the knowledge that many young geographers fresh to the subject feel unsure about the necessary physical sciences to explain atmospheric processes, more emphasis than is usual has been given to the basics, short of mathematical treatments. As an introduction to the subject, just enough bait leading to regional studies of world weather and climate has been laid to lure readers into exploring further for themselves. At the research frontiers, too, explanations for climates are sought through a greater knowledge of atmospheric processes rather than the other way around.

The first chapter enlikens the atmosphere to a machine governed by the Earth's size and fuelled by the Sun, and the second unravels how the whole restless atmosphere moves or circulates. The central chapter appropriately hinges upon the energy involved in the circulations and, especially how the transfers of moisture and heat affect us as well as the weather. Therefore, it is fitting that the fourth chapter should focus upon the processes more immediately around us. The final chapter looks outwards and glimpses into the future concerning broader global patterns.

The prospect of changing patterns and climates gives some purpose and urgency to the study of the atmosphere. The topic needs no further justification to the geographer than the testimony by the American climatologist, Reid Bryson, to no less than a committee of the United States Senate in 1974:

'The climate of the earth is changing . . . and changing in a direction that is not promising in terms of our ability to feed the world'.

Two years later the World Meteorological Organisation was moved to an unprecedented announcement warning that we were overheating the atmosphere with our activity. These are prospects that we cannot ignore or fail to study with care. The weather can be fun and a source of much pleasure too.

April 1979   JAMES D. HANWELL *Wookey Hole, Somerset*

Acknowledgements

This book owes much to several generations of sixth form students whose bemused frowns and informed nods have clarified the following introduction to meteorology and climatology. Dr Darrell Weyman encouraged it to be set down and his advice has been particularly helpful along with that also given willingly by J. N. Bates, W. J. Hayes, R. J. Scourse and J. A. Williamson. Dr Donald M. M. Thomson kindly prepared the author's own photographs and Phillip Romford provided more expertise including the bonus of a picture for the cover and Photograph 10. Dr Tim P. Burt carefully processed the data and computed Figure 4.23 from records given earlier to the author by the former Bristol Avon and Somerset River Authorities. The Controller of Her Majesty's Stationery Office gave permission to use Crown Copyright material supplied by the Director-General of the Meteorological Office for Figure 4.2 and the satellite picture Photograph 6 was provided through D. J. Causer and is reproduced by courtesy of the United States Department of Commerce, National Oceanic and Atmospheric Administration.

All concerned are acknowledged and gratefully thanked. Hopefully, they will not mind sharing this credit with a mention of the dismal weather during the summer of 1978 which offered few distractions from the task of writing the main text.

Contents

List of Tables

Chapter 1

Activity in the Atmosphere

On a globe the size of an ordinary football, the atmosphere would be represented by a skin barely thicker than a single page of this book. It is the outer layer of Earth and it 'clings' to the surface because of the pull of gravity inwards. The activity that occurs within it is vital to all life and to most features of the landscape.

The separate acts of picking up this book and reading this far require different types and amounts of effort or energy. In the atmosphere, too, each activity involves certain forms and quantities of energy. Therefore, when studying atmospheric processes it is helpful to start with a review of the energy needed to maintain different activities.

A Primary sources of energy

How to use more and more energy without harming or, perhaps, even destroying the natural environment has become one of the greatest unsolved problems facing Man. Some authorities argue that key atmospheric energy sources such as solar and wind power should be harnessed more effectively and turned directly into heat energy to meet the growing demands for cheaper power supplies.

Sources of energy take many forms which can be related to present activity in the atmosphere or traced back to such activity in the past. The former may be regarded as renewable or *primary* energy sources and the latter as non-renewable or *secondary* kinds. Typical secondary sources are found stored within the ground such as coal and oil. They may also be called 'fossil' fuels. Most renewable sources of primary energy on the other hand are part of everyday weather elements such as sunshine and the wind. So, by burning fossil fuels, the

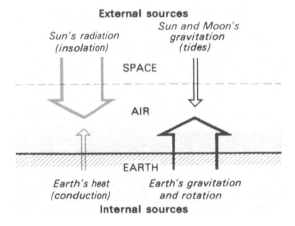

Figure 1.1 Constant sources of energy

stored energy of past activity is added to that of the present. This could be altering the world's weather in an unnatural way.

Figure 1.1 shows that the energy available to the atmosphere comes from external and internal sources. Apart from balancing each other, these both contain thermal and mechanical forms dependent upon *heat* and *mass* respectively. The Sun's **radiation,** or **insolation,** is the dominant source of heat energy while Earth's motions and **gravitation** exert most influence over masses. Continuous streams of radiation from space bathe the atmosphere, and gravitation remains a constant force internally.

Six major exchanges of heat energy and mass within the atmosphere are arrowed in Figure 1.2. They are numbered in sequence according to the distribution of energy throughout the system. Radiation energy between the atmosphere and space heads the list. This activates the transfers of

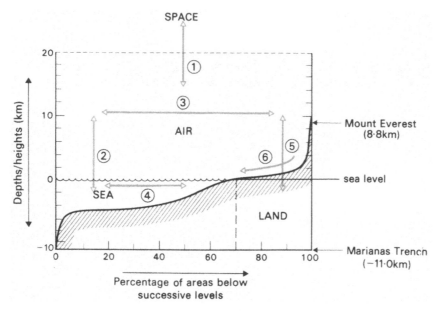

Figure 1.2 Main exchanges of energy and mass

both heat and mass energy from the seas into the air and over the land. Water is the substance chiefly involved in these intermediate processes. At the end of the chain, it is returned to the sea via rivers because gravity ultimately takes over the movement of masses. Thus, the role of water is crucial and it should be noted that the oceans cover about 70% of the Earth's surface area. The energy chain outlined here may be recognised as an integral part of the water- or **hydrological cycle.**

The continuous cycling of heat and mass energy in the atmosphere may seem contrary to the unattainable dream of a perpetual motion machine. In fact, this is not so and it is more properly called a perpetual *movement* machine because no *extra* energy is produced *within* the atmosphere, discounting the burning of fossil fuels already mentioned. Such movement results from the fine balance that has existed for so long between the output of radiation from the Sun and the overall effects of Earth's gravitation.

Briefly, **radiation** is the transfer of energy through matter or space by electric or magnetic fields suitably called **electromagnetic waves** (see Fig. 3.9). High-energy waves are emitted from the tiniest particles in the nucleus of an atom, whereas low energy is associated with larger whole atoms and molecules. Highest-energy waves are known as radioactivity since they are generated by the splitting (fission) or joining (fusion) of particles

and low energy results from the vibrating and jostling (collision) of molecules. Nuclear power stations re-create these processes to produce the electricity consumed in our homes, and the Sun may be regarded as a huge furnace in which hydrogen atoms fuse into helium at immensely high temperatures. Streams of this energy flow virtually uninterrupted across the Solar System, and Earth receives a small fraction via the atmosphere. Here it is partially absorbed by matter of increasing size, first by exciting electrons as in ionisation and then by stimulating molecular activity at lower energy levels. The latter is sensed as heat (see Ch. 3.D). This chain shows that radiation is progressively degraded or dissipated from tiny nuclear particles to bigger molecules of matter.

Gravitation is a more complex form of energy since it can be considered at both local and universal scales. Although we conceive of space and time as generally infinite, both must be ordered systematically when we study physical processes. On Earth, Newton's classic view that the space and time stage appears locally 'flat' is a good practical approximation of how terrestrial gravity works from one place to another. Throughout the Universe, however, it takes Einstein's elegant Theory of Relativity to show that gravitation is actually accomplished by a combined space–time system that is generally 'curved'. Here, the amount of curvature is proportional to the energy present in

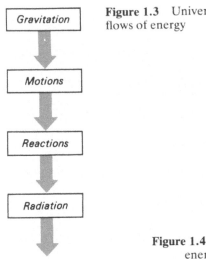

Figure 1.3 Universal forms and flows of energy

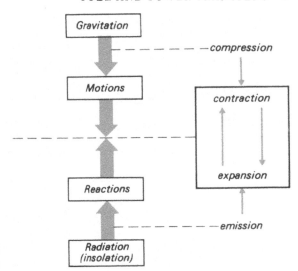

Figure 1.4 Forms and flows of energy in the atmosphere

that part of space at a given time. The impact of this theory on physics compares with that on geographical discovery when the Earth was shown to be spherical rather than flat. In short, gravitation is greater with bigger masses and more ordered systems (see Ch. 2.D). It must not be simply taken for granted in studying the atmosphere.

Figures 1.3 and 1.4 show the difference between flows of energy in the Universe and on Earth respectively. Figure 1.3 illustrates that the total available energy within the Universe is not destroyed but becomes more disordered or chaotic. This means that low-grade energy cannot be used without the help of higher-grade forms. Figure 1.4 shows how this important general principle applies in the special case of the Solar System and Earth in particular: radiation from the Sun and the cycling of heat and mass energy in the atmosphere are assisted by gravitation as outlined below.

First, a huge body such as the Sun exerts an enormous gravitational field which not only assists the fusion process generating radiation from its surface but also controls the orbits of distant planets such as the Earth. Secondly, terrestrial gravity attracts greater crowds of air molecules at the bottom of the atmosphere to increase the chances of more collisions there. Variations of this activity in both space and time on Earth lead to stronger heat- and mass cycling nearer the ground and ocean surfaces rather than aloft. In active or high-energy regions the air expands, whereas it contracts in less active or low-energy regions.

Solar radiation counters the effects of Earth's gravitation and 'holds up' the atmosphere. In effect, they are playing a vertical tug-of-war with air columns as the ropes. Fortunately, the ultimate catastrophe of a total victory never occurs: gravitation remains the reigning champion and radiation the ever hopeful contender. If this radiation ceased, gravity would take over and the whole atmosphere would collapse!

B Fuel and power transfers

The atmosphere's activity is like that of a giant engine continuously fuelled by the Sun. As with an engine, the action of any part may be linked to the total power output for the machine as a whole. These depend upon the size and mass of the parts concerned and, of course, the time they spend 'running'. In other words, the three fundamental physical quantities of **length, mass** and **time** are all involved. Together, these quantities can be used to define and measure the **power** produced or generated. Table 1.1 lists ten steps in order to show how the definition of power is built up from the three fundamentals.

It is emphasised that energy is strictly the capacity or capability to do work. In fact, it can exist whether or not work is actually being done. Unused energy, like foods and fuels, is actually stored as **potential** energy. Alternatively, as with a body of air which could sink earthwards under the influence of gravity, the potential is attributable to

Table 1.1   Ten steps to power

Quantities	Definitions	Examples
1 Area	length×breadth	a patch of the ground surface
2 Volume	area×height	size of a solid body
3 Density	mass per volume	matter in a body
4 Velocity	length per time	speed in a direction
5 Acceleration	velocity per time	change of speed in a direction
6 Momentum	mass×velocity	impetus of a moving body
7 Force	mass×acceleration	strength in a direction
8 Pressure	force per area	strength over a surface
9 Work	force×length	effort for a distance moved
10 Power	work per time	flow of energy

position. Once at work, however, the energy becomes **kinetic** giving rise to motions and reactions. The activity may be mechanical, chemical, thermal or even electrical. One can be converted to another as in a food chain when solar radiation is converted into chemical energy by plants, which is then consumed by animals to produce heat and mechanical activity.

Conversions or transformations of energy from one form to another are governed by two general laws: the first holds that the total amount of energy remains constant or is *conserved*; and the second holds that the energy must flow towards the direction in which its activity becomes more chaotic or *random*. As mentioned earlier, such activity is both thermal and mechanical, and so the pair are aptly referred to as the **First and Second Laws of Thermodynamics.**

Figure 1.5 illustrates the overall transfers of potential to kinetic energy within the air, Earth and space systems. Solar radiation assisted by gravity generates the power to drive atmospheric processes over a period of time and ends up as heat and mass energy transfers. During the same time, the Earth itself radiates an equivalent amount of energy to space so that the gains and losses balance; otherwise, the atmospheric machine would overheat or freeze up.

Both potential and kinetic energy are measured in units called **joules** (J). Every 4·18 J is equivalent to a calorie or, more correctly, a gramme-calorie, since its standard is the heat energy required to raise the temperature of a gramme of water through a degree Celsius (°C) under controlled conditions. Various other equivalents are commonly used by climatologists and the calorie unit used by dieticians is actually the kilogramme-calorie.

Joules and gramme-calories are common measures of activity and food values. 10^6J, or nearly 2·4 million gramme-calories represents the daily energy intake and output of a moderately active adult person. Since this amount is needed every

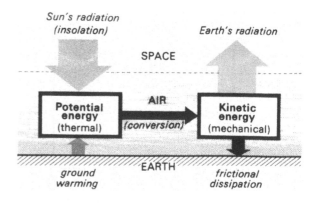

Figure 1.5   Potential and kinetic energy

day, then the continuous energy requirement or *flow* for an active person averages almost 100 J every second. As defined in Table 1.1, this value will be the power rating for a man.

Since a joule every second is the same as the absolute power unit called a **watt** (W), basic man-power may be thought of as about 100 W. By comparison, therefore, a man is hardly on par with the power rating of an ordinary domestic electric light bulb! In fact, depending upon an individual's size, the basic body functions tick over between 65 W and 85 W. They rise to about 400 W during a brisk walk and might peak to 700 W with violent exercise.

The most recent measurements of the intensity of solar radiation taken at the outer edge of the atmosphere from space craft give a standard value of 1353 W over every square metre facing the Sun. Because this figure varies little over long periods of time, it is known as the **solar constant.** At any time, the part of the Earth's atmosphere facing the Sun receives about 10 000 million times more energy than sustains a single person on the ground. Yet, we shall see that only a small fraction of this immense flow is put to work in the atmosphere or, more particularly, at the Earth's surface.

C Power supplies and balances

Nowadays, the power consumption of all appliances such as heaters, record players and so on are rated in watts, or a multiple such as the kilowatt (kW). Being in essence a machine, the atmosphere is no exception and it is convenient to measure its flows of power in vast units of 10^{12} W, this being a million times a million watts (a unit called the terawatt).

Figure 1.6 illustrates the streams of energy that power the entire atmosphere and, indeed, the biosphere. Each flow is a fraction of the solar constant and there is a general global balance between inputs and outputs as explained already. These streams are considered in more detail in subsequent chapters, and Figure 1.6 is worth comparing with Figure 3.13 later. Suffice to point out at this stage that all weather and life on Earth are sustained by tiny percentages of the total power available. Neither the atmosphere nor Man are particularly efficient at utilising the available power.

Coincidentally, and perhaps significantly, the basic food needed by the rapidly growing world population runs at 10^{11} W. If the extra power for industrial activity and growth is added, then future

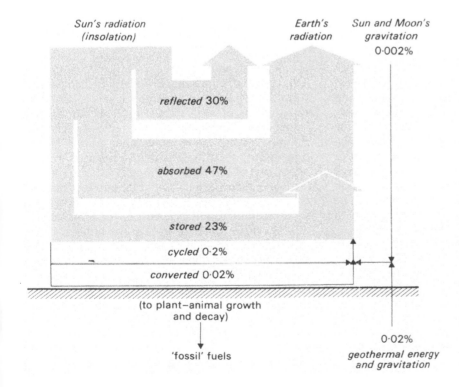

Figure 1.6 Streams of energy

Table 1.2   Continuous power supplies and current needs by the world population

Continuous natural power supplies (10^{12} watts)		Current human power needs (10^{12} watts)	
absorbed by atmosphere and converted to heat	81 000 (47%)	consumption of oil reserves	6·1 (54%)
stored in hydrological cycle, oceans and ice	40 000 (23%)	consumption of coal reserves	2·4 (22%)
used to drive circulations of winds, waves, etc.	370 (0·2%)	consumption of gas reserves	1·2 (11%)
used in reactions and all organic growth	40 (0·02%)	generation of hydro electric power	0·7 (6%)
internal geothermal and gravitational energy	32 (0·02%)	generation of nuclear power	0·5 (4%)
external tidal energy from Moon's gravity	3 (0·002%)	basic food needs by human population	0·3 (3%)

Note: Approximate percentages of totals are given in brackets. The remaining 30% of the 'supplies' has been reflected and unused.

total human energy requirements could approach the same order of magnitude as atmospheric activity at 10^{12} W. Table 1.2 itemises the continuous natural power supplies and present needs of the world population for comparison. It shows that the atmosphere takes in about 70% of the solar radiation to its fringe. This amount then splits up or cascades throughout the atmosphere so that each activity gets a share. Although existing human needs of 11·2 terawatts in total may appear to be relatively puny, future increases would begin to rival the critical requirements for plant growth.

As fossil fuels run out, it will become necessary to tap more renewable sources of primary power. With interference on a massive scale, the atmosphere would be the first to suffer. In Chapter 4.E, it will be seen that atmospheric processes have intensified over large urban areas, and a better understanding of this problem is regarded as urgent by many environmental scientists studying climatic changes globally.

D   Scales of activity

The range of atmospheric activity to be spanned from the smallest to the largest levels is immense. Illustrations of such ranges may be based upon broad orders of magnitude. Examples are given in Figures 1.7 and 1.8 as divided bar graphs. Notice how in both the divisions jump various powers or exponents of ten rather than in regular units.

It was seen earlier that different sources and chains of energy influence bodies of different sizes; so, atmospheric activity is best divided on the basis of **scale**. Figure 1.7 shows that the familiar quantities of distance and time are good pointers to five divisions centred upon the atmosphere's thickness and one complete revolution of the Earth upon its axis. These are matched in Figure 1.8 with divisions of mass and power centred upon those of one man. On both diagrams, the divisions are guides to identify four scale bands for studying atmospheric activity, viz. global, synoptic (zones covered by weather maps), meso-(intermediate regions) and

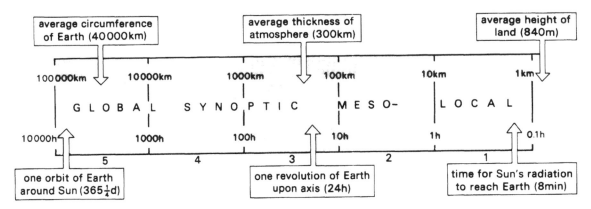

Figure 1.7 Scales of activity

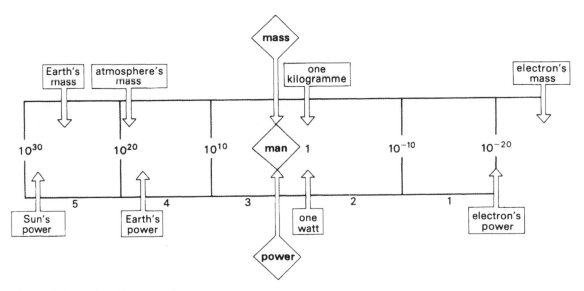

Figure 1.8 Scales of mass and power

local areas. These are also named in Figure 1.7. The **global** scale includes the long-lived general circulations throughout the atmosphere and the **local** one focuses upon the short-lived activity near the ground at different places. Between them lies transitional activity covering zones and regions at continental and oceanic scales, the most important being the **synoptic** one as the main link to either extreme.

Figure 1.9 shows that more energy is involved with the global and local scales because of planetary and particle activity respectively. Global requirements exceed those of all smaller scales combined. A noticeable 'slack' or gap is evident at the intermediate scales and this represents the general mixing of air where the effects of radiation

and gravitation are best matched. The gap is less within the Tropics because global and local processes are more closely linked nearer the Equator. At

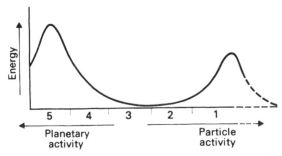

Figure 1.9 Energy at different scales in the atmosphere

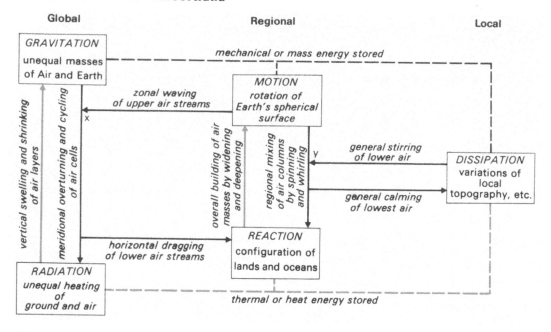

Figure 1.10   Activity in the atmosphere

higher latitudes such as the temperate regions, however, the gap is very clear. The reasons for this will be considered later. It is sufficient to note here that the most intense activity occurs within the Tropics, which represents the 'boiler room' of the atmosphere.

Although it is convenient to separate different members of the atmospheric machine for study on the basis of scale, in reality the smaller parts are incorporated within the larger ones as 'wheels within wheels'.

E   The atmospheric machine at work

The 'wheels' of the atmospheric machine are depicted in Figure 1.10. Four main components interlock or engage the global, synoptic, meso- and local scales of activity.

On the left side of Figure 1.10, the expansion and contraction of air causing the cycling of heat and mass energy represents the engine and *driving* parts. These are found at most scales; for example, from tropical to temperate latitudes, between warmed regions and adjacent cooled ones and so on. In effect, such activity resembles that of huge heat pumps kept at work by the unequal receipt and distribution of energy within the atmosphere.

In the centre and to the right side of Figure 1.10,

the activity summarised involves different ways in which the energy flows from the heat pumps (the sources) to the cold regions (the sinks). Here the spinning surface of the Earth exerts its greatest influence; the flows are deflected and dragged into great waves and whorls remarkably similar to those made when stirring bath water to mix the hot and cold supplies more effectively. Lastly, they are dissipated as heat and moisture to space and the ground respectively.

The points labelled x and y on Figure 1.10 indicate where the 'wheels' at the different scales engage. Each one creates 'hold-ups' as kinetic energy is being expended and this is manifest as **turbulence.** Such activity seems to generate different types of weather. Therefore, the general layout of Figure 1.10 is also a convenient guide to the following chapters on motions, then moisture and heat (reactions) and so local processes.

The basic activity of unequal heating by the Sun and unending spinning of the Earth lead to the continuous sharing out of the available energy within the atmosphere. Finally, this situation may be put in its most telling way: if the Sun were to be suddenly switched off, then Earth would have to rely solely upon its own resources. In these circumstances, there is barely enough fuel to keep the machine running for a week!

Chapter 2

Motions in the Atmosphere

The restless nature of the atmosphere is its most obvious characteristic and undoubtedly the first aspect to arouse Man's curiosity. By investigating atmospheric motions first, too, we acknowledge this here before delving deeper into their causes in Chapter 3. In reversing the usual order of taking radiation first, we follow the same historical path that led researchers to a basic understanding of air motions long before the more complex exchanges of radiation energy were unravelled.

Motion is the movement of any body from one place to another. The body may be a person, a minute particle such as a molecule, or a huge mass of air covering a whole country. On the ground, movements occur along established routes such as roads and rivers, which are readily mapped. In the atmosphere, however, there is far greater freedom of movement and bodies of air cannot be seen anyway.

A Methods of approach

There are two ways of tackling the problem of studying atmospheric motions. Once a body of air has been located, it may either be tracked in relation to a framework of fixed points on the ground or actually followed airborne so that changes can be examined en route. The first is akin to a series of snapshots as in cartoon films, each 'still' or frame representing an instant in time. Therefore, it is called the **synoptic** approach. On the other hand, the second approach is said to be **dynamic** because it deals directly with the motions 'in flight'.

Maps are basic to both approaches outlined above; however, the dynamic approach relies more upon known mathematical equations of motions. These express the principles of the conservation of momentum, mass and energy numerically.

In addition to the basic assumption that the atmosphere 'clings' to the Earth and is vertically balanced, four fundamental equations are used to solve the four unknown variables of temperature, pressure, density and velocity which account for the air's restless behaviour (see Ch. 2.E).

The synoptic and dynamic approaches both reveal different aspects of the patterns of motions and the processes which cause them, and so each supports the other.

Although geographers often avoid studying global motions, the purpose of this chapter is to describe and explain the **general circulation.** In much the same way that the theory of plate tectonics has unified the study of geology, so the general circulation is the central problem in meteorology. It is the hen without which it is difficult to account for the egg! As geographers rather than meteorologists, however, the conundrum of which should come first is hopefully answered by the greater importance attached to the effects of different surfaces from place to place in subsequent chapters once the basic processes have been described. Whereas the model of the general circulation is the goal of most studies in meteorology, it is the best point to start for geographers.

Emphasis is given to synoptic methods in the following account and key dynamic principles are raised only where necessary. At the risk of over-simplifying what is actually complex, therefore, explanations stop short of mathematical treatments. The plan now followed first defines the bodies of air and the way they are mapped before grappling with how and why they circulate. The consequences of air moving within yet more air are then examined as a basis for studying the effects of the spinning Earth.

B Bodies of air

A body is something distinct from its surroundings and is, therefore, discrete or individual. Isolated clouds and puffs of smoke testify to the presence of discrete bodies within the air. On the whole, the quicker these develop or disperse, the more vigorous are the motions involved.

Bodies are best described initially by their *shape* or form and then their *size* (Table 2.1). Once again, clouds and smoke are good visual indications of these characteristics, particularly that of form. Table 2.1 starts with separate localised bodies and finishes with broader grouped patterns; examples are cited by references to photographs in later chapters.

Figure 2.1 shows how the five different forms of air are related to the main motions at the global, intermediate or regional, and local scales (see Fig. 1.7). Notice that northern hemisphere situations have been chosen and that both smaller areas are shown as 'bites' from the larger ones.

We shall see in this chapter how the *global* circulation arises because warm air is forced aloft in the Tropics. Once at high level, the warm air spreads polewards gradually turning into strong upper westerlies owing to the spinning surface of the Earth. Meanwhile, sinking cold air pushes towards the Equator at low levels from the polar regions and slows down. The resultant 'pile-up' of air layers at all levels in mid-latitudes is partly resolved by the overturning of huge tropical and polar air **cells** and partly by mixing. Large masses of warm air break loose into higher latitudes and cold ones push even further towards the Tropics. Figure 2.1 illustrates the overall flows with respect to the bodies outlined in Table 2.1.

The regional mixing that occurs in mid-latitudes is also shown in the middle diagram of Figure 2.1. Great **waves** in the upper westerlies seem to develop when strong air streams off the Pacific are forced to soar above high mountains such as the Rockies of North America in their path and then dive across the following interior plains. Giant **vortices** (see Table 2.1) are set up by these waves; a soaring flow pulls up cores of cold air as in mid-latitude **cyclones,** and a diving one pushes down cores of warm air in **anticyclonic** systems. The British Isles takes the brunt of such systems as they are driven across the North Atlantic into Europe. These largely mechanical mid-latitude or temperate systems will be seen to contrast with the ther-

Table 2.1 Types of air body

Visual forms	General descriptions	Typical examples
1. Cells (see Photo. 1)	individual, often spheroidal, parcels of varying size from bubbles to large air masses	local cumulus or heaped clouds, billowing fogs, mists, smoke puffs and palls
2. Columns (see Photo. 2)	vertical piles, towers of 'chimneys' of air of great height but restricted width	colossal cumulus or congested clouds, 'boiling' fogs and smoke pillars
3. Layers (see Photo. 3)	horizontal beds, sheets or 'strata' of air of limited thickness but great extent	general stratus or layered clouds, lowlying fog blankets, smogs and smoke screens
4. Bands (see Photos 4 and 5)	textural air streams in long parallel lines or meandering waves with definite margins	filamental cirrus or veiled clouds, often streaming and wisped, and long smoke plumes
5. Clusters (see Photo. 6)	spiral or gyral eddies of spinning air with cores which have either updraughts or downdraughts	radial whorls of cloud bands around vortices called cyclones and anticyclones; also whirl winds and dust devils

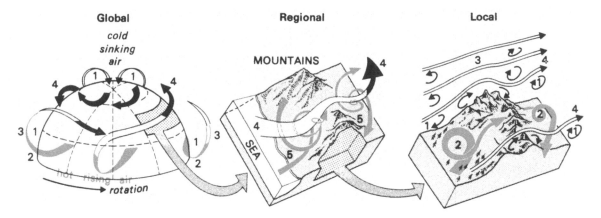

Figure 2.1 Global, regional and local motions in the air

mal ones more common to Tropical weather (see Table 2.4 and Ch. 5).

At the local scale shown on Figure 2.1, terrain plays an important role by causing further turbulence and mixing. Unequal warming of slopes and the rolling or rotary action of surface winds over uplands create the distinctive updraughts and downdraughts which are so much a feature of local weather.

Thus, as concluded in the first chapter (see Fig. 1.10), the bodies of air move about the atmosphere striving to distribute their energy uniformly over the unequally heated surface of the Earth.

C Maps of the air

As there are few visible outlines to bodies of air, other properties are needed to plot their shape and size onto maps. Just as relief and rivers, settlements and communications and so on are basic to maps on the physical and human geography of the land, so six primary elements are measured to prepare maps of the air. Thus, this section may be subdivided into two: first, mapping the elements, and secondly, using air pressure maps.

Mapping weather elements
Table 2.2 lists the six elements. Sunshine is included with radiation since it is the most obvious and easiest portion of solar radiation to measure. Temperature and density are linked because as warmed air swells or expands so fewer molecules occupy a given space relative to cooled air which shrinks or contracts. Also, since winds are the result of air pressure differences, they are not considered as a separate element. Except for cer-

tain rather special optical and chemical characteristics such as rainbows, mirages, airborne dust or aerosols and manmade fumes, the great variety of everyday weather derives from combinations of the six primary elements.

The instruments that measure the primary elements can only record values at a particular place and instant or short period of time. Such observations are called **point** values, with regard to both space and time. Repeated or even continuous readings from *one* instrument will show how an element alters with time, of course, but a **network** of stations is required to map distributions in space.

Over 7000 permanently manned stations comprise the network by which the various agencies of the World Meteorological Organisation (WMO) observes global processes. About 4000 of these fulfil WMO requests for surface observations at standard times every 6 hours. Of these, only some 800 also take regular upper air readings because the necessary balloons and rockets to do this are costly. To a great extent, the lack of the latter is offset by photographs taken from weather satellites (see Photo. 6). Meteorologists have enjoyed the unprecedented advantage of almost continuous views of atmospheric motions from space since the first weather satellite was launched, somewhat boldly, on 1 April 1960. Nowadays, the *direct* measurements from the ground and the *remote* ones from space are complementary. The former provide much needed 'benchmarks' to interpret satellite photographs.

Fortunately, all the primary elements except clouds and precipitation are continuously distributed throughout the atmosphere. This makes it

Table 2.2 Primary elements of the atmosphere

Elements	Instruments	Measurements	Maps
radiation and	radiometers	power per unit area or flux	isolines
sunshine	sunshine recorders	duration of bright sun	isohels
cloud cover	estimation	fraction of visible sky covered (eighths)	isonephs
temperature and	thermometers	degree of sensible heat energy	isotherms
density		mass per volume of air	isoteres
pressure and	barometers	force or weight per area at any height	isobars
wind (direction)	anemometers (and vanes)	velocity of motion (direction of flow)	isotachs (isogons)
humidity	hygrometers or psychrometers	amount of water vapour per amount of air by either mass, volume or pressure	isolines or isobars
precipitation	rain gauges	depth of rain in a given time or its equivalent from snow, etc.	isohyets

possible to map them with equal or constant value lines called **isopleths** or, simply, isolines. These fill in the 'dead' ground between stations for which no actual readings exist. For convenience, too, even clouds and precipitation are best mapped in this way and errors arising from discontinuous distributions are ignored.

Isopleth maps are commonplace in atlases and several follow in this book. To distinguish each version, it is usual to follow the prefix *iso-* (which means equal in Greek) with a term (also of Greek origin) that defines the particular quantity measured. For example, another name for the familiar contour line is 'isohypse'. There are many features in common to maps of the land and air in this respect. Indeed, isopleths were used on climate maps long before contours were used to depict surface relief.

Using air pressure maps
Since the middle of the last century, air **pressure maps** have formed the basis of plotting changing weather systems. Appropriately, air pressure and density are the only two primary elements which are not really weather characteristics in themselves

but the *causes* of motion. Movement occurs along **pressure gradients** from high to low values. To the meteorologist, therefore, isobar maps give similar information to contour maps regarding relief gradients. Such maps are televised several times every day and are published in some daily newspapers to explain weather forecasts. In fact, they must be the most common type of map encountered by the general public.

In the same way that contours must be related to a datum such as mean sea level, isobar maps are also drawn at particular altitudes or levels. Mean sea level or surface charts are the obvious *base* maps for the atmosphere too. However, similar maps can be drawn for any height.

Meteorologists are confronted with a layer of the atmosphere (at least 20 km thick) in their 6-hourly quest for details of motions. The same penetration into the Earth, let alone its frequency, would be well below the depths of interest to most geologists and certainly the geomorphologist! A branch of geography in which the basic maps need to be redrawn at least four times a day may seem daunting enough. The great effort and expense to do this are indicative of the importance attached to

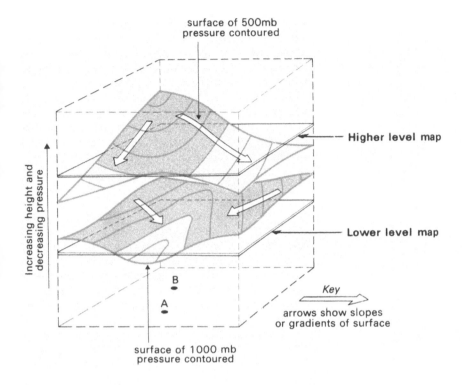

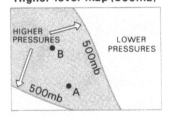

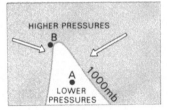

Figure 2.2 Pressure surfaces and maps at different heights

monitoring the continually changing patterns of atmospheric motions.

The following synoptic maps or charts form the basis of analysing the changing patterns 'in depth' To them might be added the weather satellite photographs which make detailed soundings from vantage points in space.

(a) *Sea-level charts* (see Fig. 4.2). Based upon isobar maps of simultaneous air pressure records corrected to mean sea level from a station network, e.g. the **Daily Weather Report** of the British Meteorological Office.

(b) *Upper-level charts.* Based upon height records of selected isobars from tracking balloons and rockets sent up from several stations simultaneously, e.g. the **Daily Aerological Record** of the British Meteorological Office.

Figure 2.2 illustrates how isobar maps are drawn at two altitudes or levels, one above the other. A block of the atmosphere is shown with the undulating surfaces where the heights of pressures at 500 millibars (mb) and 1000 mb were recorded. Both surfaces have been contoured in the drawing. The resultant pressure maps sliced through each surface at appropriate levels are also given in Figure 2.2. Notice that the 500 mb map reveals an upper high pressure spur or ridge while the 1000 mb one shows a valley-like depression or trough.

In the case just described, an upper ridge is seen above a lower trough because the air layer between the two isobaric surfaces has swollen in thickness above points A and B on the ground. Indeed, **thickness charts,** which use isopleths to plot amounts of swelling or thinning of layers between given **isobaric surfaces,** are yet another way of mapping air bodies. Figure 2.3 is a simple example derived from the situation illustrated above. Such maps are prepared by plotting the vertical differences between crossing contours when the two surfaces are superimposed.

As mentioned at the beginning of this section, because expansion, temperature and density are directly related, thickness maps are guides to mean temperatures of air layers. In Figures 2.2 and 2.3, a warmer swollen air cell lies above A and B between the two pressure surfaces. Such cells are stores of energy and give rise to the *horizontal* forces that start the air moving.

Figure 2.4 shows how this is produced. The three vertical blocks depict columns of air at different temperatures and correspondingly different pressure changes with height. At the level chosen, therefore, a horizontal **pressure gradient force** exists (called **P**) to drive bodies of air in the direction shown. This would be apparent as a wind, of course.

This section may be concluded by reassuring ourselves that most terms used to describe pressure maps are shared with those which name relief

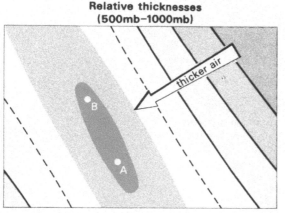

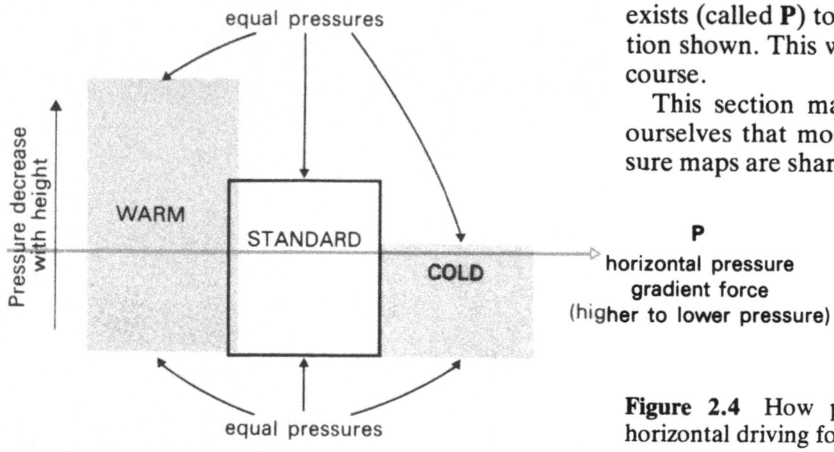

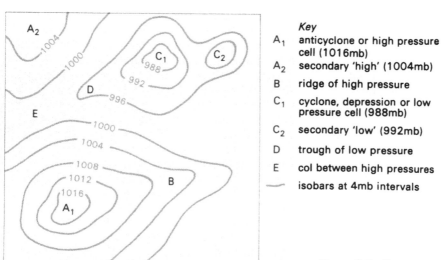

Relative thicknesses
(500mb–1000mb)
thicker air
B
A

Figure 2.3 Maps of the thicknesses of air layers between pressure surfaces

equal pressures
Pressure decrease with height
WARM
STANDARD
COLD
P
horizontal pressure gradient force
(higher to lower pressure)
equal pressures

Figure 2.4 How pressure differences give rise to horizontal driving forces

A2
1004
1000
C1
988
C2
992
D
996
E
1000
1004
1008
1012
B
1016
A1
Key
A1 anticyclone or high pressure cell (1016mb)
A2 secondary 'high' (1004mb)
B ridge of high pressure
C1 cyclone, depression or low pressure cell (988mb)
C2 secondary 'low' (992mb)
D trough of low pressure
E col between high pressures
isobars at 4mb intervals

Figure 2.5 Pressure systems on isobar maps

features on contour maps. Figure 2.5 is a composite surface pressure chart illustrating examples of the main terms. Just as steep slopes from hills to valleys are shown by contours which are closely spaced horizontally on relief maps, so isobars that are close together show strong *horizontal* gradients from high to low pressure regions on weather maps. Likewise, maximum gradients cross the isobars at right angles.

D Circulating air

The *vertical* circling, and so overturning, of air cells is called **convection**. The horizontal components of such circulations may be distinguished as **advection**. Convection and, therefore, advection are the result of unequal warming of the air from below. In this section the broad reasons for global convection will be reviewed first and then a general explanation for the vigour of such circulations follows.

Global convection
We need go no further than a heated room at home to appreciate how convection causes air to circulate. Figure 2.6 gives a sequence of diagrams with explanations outlining the stages involved. Imagine that they represent a room which has heaters on the left side and large windows on the right. Air rises above the warm source and subsides over the cool sink. The resultant overturning cell will circulate more vigorously with a bigger temperature difference between the source and sink; this could be achieved by either increasing

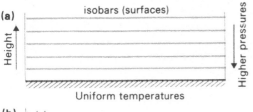

(a) Layered air above uniformly heated surface. Regular decrease of air pressure with height and no air motions.

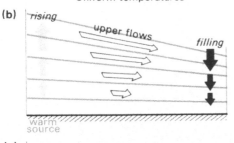

(b) Expansion of layers above warm source causes rising and increase in potential energy. Upper flows of kinetic energy develop down 'slopes' to fill and depress other side. Motions begin as in (c) or (d).

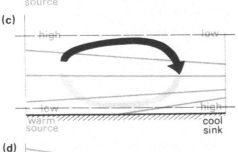

(c) Motions and overturning along pressure gradients. Filling and depression above cool sink lead to increase in surface pressure corresponding to surface low above warm source. Note that pressures are relative at the two separate levels shown.

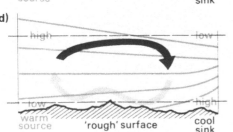

(d) Surface 'drag' or friction of lower flow is countered by increase in pressure above cool sink and decrease above warm source to 'drive' flow faster over rough features. Hence, stronger surface pressure gradient set up.

Figure 2.6 How air circulates

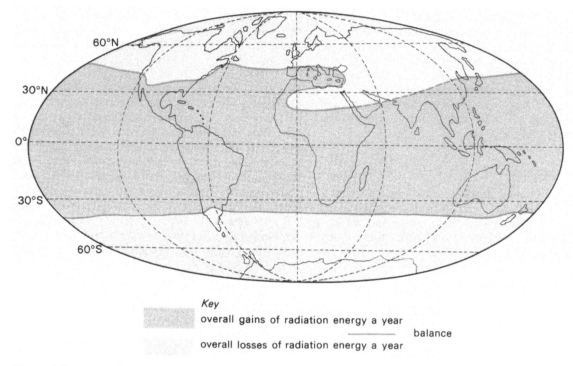

Key

overall gains of radiation energy a year

balance

overall losses of radiation energy a year

Figure 2.7   Net radiation balances in the atmosphere

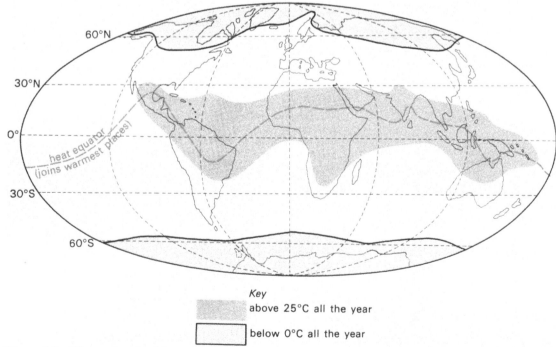

Key

above 25°C all the year

below 0°C all the year

Figure 2.8   Warmest and coldest regions

the supply of heat energy or withdrawing more energy from the sink, e.g. turning up the heater or opening the window. This simple explanation of circulating air is called the **advective model**. Synoptic-scale pressure systems associated with such circulations are illustrated in Figure 2.21 and explained in Chapter 2.F. Local examples such as land and sea breezes are given in Chapter 4.C (see Fig. 4.13).

If the surface beneath the overturning cell is uniformly smooth, then the convection may be termed as **free**. Alternatively, with irregular surfaces, pressure adaptations occur so that the low-level advective element in particular becomes **forced**. The latter situation is also shown on Figure 2.6. Houses can be designed to take advantage of both forms of convection for circulating warm air or ventilating rooms. These methods of heating and air conditioning are called thermosiphoning. So, the proposition at the end of the first chapter that air is pumped around at all scales is borne out as being caused largely by convection.

The existence of large-scale global convection was postulated by George Hadley as long ago as 1735. In the next century, William Ferrel made the first important study of the mid-latitude westerlies and then, in 1889, argued that smaller-scale cyclones resulted from localised convection above strongly heated places. Both realised that the Tropics were warmer because they received more intense solar energy than regions polewards. In the absence of a theory for radiation exchanges, however, no one at that time could explain the long-accepted pattern of greater heating in lower latitudes. Like the pioneers, we too can be content that this is so for the moment in order to concentrate upon the resultant motions. So, until explained in the next chapter, Figure 2.7 serves to show the broad latitudinal bands (**zones**) experiencing gains and losses of radiation energy. This should be compared with Figure 2.8 which gives the related temperature regions generally understood by Hadley and others. The warmest places on Earth do not necessarily coincide with the geographical Equator, but join as an isoline simply called the **heat equator**. Its mean location for the year is seen to 'follow' the more extensive land areas which heat up most in summer. The seasonal movements of the heat equator are considered in Chapters 3.C and 5.A.

Figure 2.9 illustrates a classic, if rather oversimplified, three-cell model of the general circulation along lines of longitude (**meridians**). It acknow-

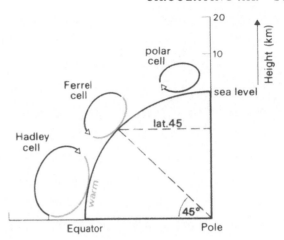

Figure 2.9 Simplified circulations from Equator to Pole

ledges the work of the pioneers by naming the dominant tropical circulation after Hadley and the mid-latitude one after Ferrel. 'Mirror image' circulations figure in the opposite hemisphere. Although this model assumes that the Earth is stationary with a uniformly smooth surface, it reveals the existence of rising air in equatorial zones, subsiding air in mid-latitude zones and further uplifts at higher latitudes (compare Fig. 2.9 with Fig. 2.1).

Related surface pressure regions are generally shown on Figure 2.10. Their most prominent features are the persistent mid-latitude high-pressure zones and tropical lows over the more heated land masses in summer. The former are established where air subsides in the mid-latitude pile-up mentioned earlier and are virtually permanent over the cooler oceans there (see p. 72). In the Tropics, the rising air drives the Hadley cell circulation.

The quartet of world maps is completed with Figure 2.11. This last one shows that the wettest and driest regions are also broadly zonal, being closely linked to the cyclonic and anticyclonic circulations respectively. More detailed relationships will emerge in later chapters.

General vigour of circulations
A general theory to explain the vigour of circulations anywhere was advanced by Vilhelm Bjerknes in 1898. It uses the key physical variables of pressure, temperature and density to explain the state of the air at any time or place.

Figure 2.12 shows three possible versions in which the relationships between isobars and isoteres (see Table 2.2) describe the strength of circu-

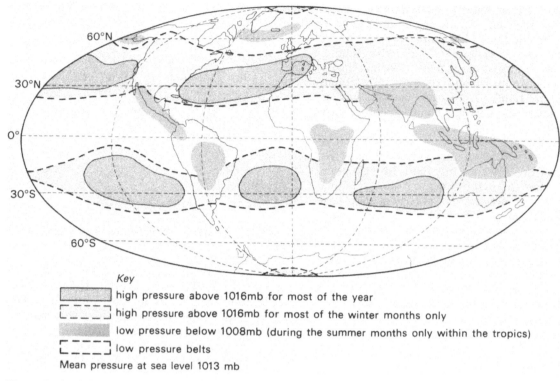

Key

high pressure above 1016mb for most of the year

high pressure above 1016mb for most of the winter months only

low pressure below 1008mb (during the summer months only within the tropics)

low pressure belts

Mean pressure at sea level 1013 mb

Figure 2.10   Highest and lowest surface pressure regions

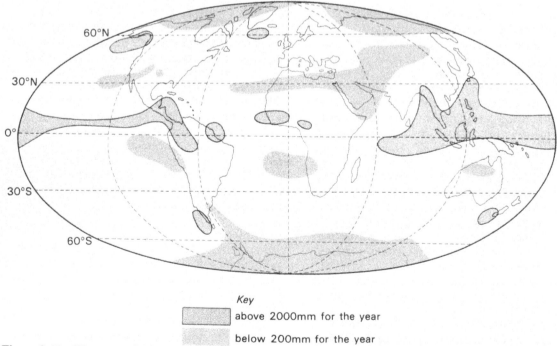

Key

above 2000mm for the year

below 200mm for the year

Figure 2.11   Wettest and driest regions

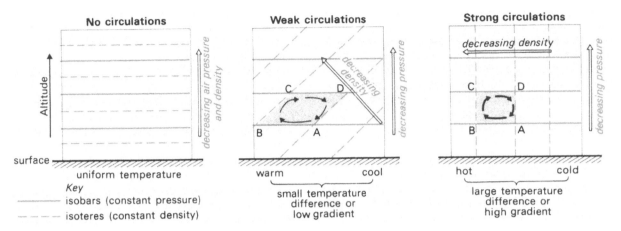

Figure 2.12 Vigour of circulations

lations. The first situation is somewhat unreal and it models a **standard atmosphere**. It is also called a **barotropic** one (which means 'circling') since both isolines are parallel to the curved surface of the Earth. By contrast, the two **baroclinic** (or 'bending') models show sloping beds of discrete air rather than uniform layers. Isobars could also be slightly inclined in the latter as shown in the larger viewpoints on Figure 2.6. Each bed becomes like a tube in a baroclinic situation and is called a **solenoid** (one is labelled A, B, C and D on the appropriate diagrams). Within every solenoid, air circulates from the corner with the higher pressure and density to the lower pressure and density ones in turn as shown (also, compare this with Fig. 2.6). The vigour of these circulations increases with the angle made between both isolines because the pressure and density differences around such a solenoid are greater. It becomes a maximum when the isolines intersect at right angles, as shown in the third diagram.

The **Bjerknes Theory** is also used to identify large **air masses** with widespread uniformity of pressure, temperature and other related characteristics horizontally, even though these vary vertically. Such situations tend to be more barotropic. The boundaries or discontinuities between different adjacent masses known as **fronts,** however, can be regarded as narrow 'walls' with strong baroclinic characteristics. This explains the greater activity along fronts and the more intense weather generated by them (see Ch. 4.A). In short, large air masses are barotropic and fronts are baroclinic. The former are closely associated with the stable high-pressure cells or anticyclones and the latter

more with mid-latitude cyclones or low-pressure systems, especially those frequenting the British Isles. More details about air masses and fronts are given in Chapter 3.E and Chapter 4.A.

Sinking air resulting from the pile-up in the general circulation at mid-latitudes (see Fig. 2.1) is largely responsible for the stable anticyclones there. The spinning surface of the Earth acts like disc brakes on atmospheric motions here because the barotropic state cannot generate sufficient kinetic energy to overcome the friction of the ground. It is no wonder that the more long-lived anticyclones tend to linger above the smoother ocean surfaces for mechanical reasons as well as the thermal ones mentioned earlier (see Fig. 2.10).

Indeed, we may ponder upon the fact that a totally layered or stratified barotropic atmosphere would soon grind to a halt and there would be no more to discuss in this book! Put another way: where there are no circulations, there cannot be any weather.

E Air on the move

Having established that convection can circulate air either slowly or vigorously, we must now consider the consequences of such movements. After all, the atmosphere is a *continuous* medium and its motions are really bodies of air moving within yet more air. This situation totally contrasts with other flows studied in geography which invariably consist of one type of material moving within or along another of a different nature. The solutions are to be found in the dynamic approach mentioned at the start of this chapter.

At the fundamental level, five key physical principles give a picture of what happens to air on the move. Each usually takes the form of an equation: the first accounts for the whole setting which constrains all motions, and the rest are rules to explain the resultant processes concerned. The equations are considered in the following order: the hydrostatic equation; the equation of state; the thermodynamic equation; the equation of continuity, and the equation of motion.

The hydrostatic equation

The **hydrostatic equation** may be thought of as the basic assumption about the atmosphere. As already seen, the entire mass of air would collapse if gravity ruled. Alternatively, if there were no gravity, everything would fly off into space. The 'pull' of gravity upwards decreases with the *square* of altitude because the Earth is spherical. More air molecules are attracted earthwards to increase pressure in the lower layers of the atmosphere. Thus, a corresponding decrease in air pressure with height provides the necessary balancing force upwards. By attempting to accelerate columns of air upwards, this pressure-gradient force makes

gravity act like a giant spring anchored to the ground. A fundamental vertical balance exists and, since the moisture in the air allows it to be regarded as a fluid (see Ch. 3.C), the equation is called a **hydrostatic** one.

The outcome of the overall hydrostatic balance is shown in Figure 2.13 which depicts the so-called **standard atmosphere.** It is based upon the resultant distributions of pressure, temperature and density (hence mass) in a vertically balanced system. This approximates to a barotropic state. Nearer the ground, the air layers get closer together and the vertical gradients stronger. It is noticeable that only shallow layers at the base and roof of the atmosphere have temperatures above freezing point. Also, it can be seen that alternate cooling and warming with height defines four main layers known as the troposphere, stratosphere, mesosphere and thermosphere. Appropriate isotherms are given as useful guides to the relatively sharp boundaries or pauses dividing these layers. Figure 2.14 gives greater details of the pressure levels within the standard troposphere and indicates with red arrows those isobars whose exact heights are measured by weather balloons

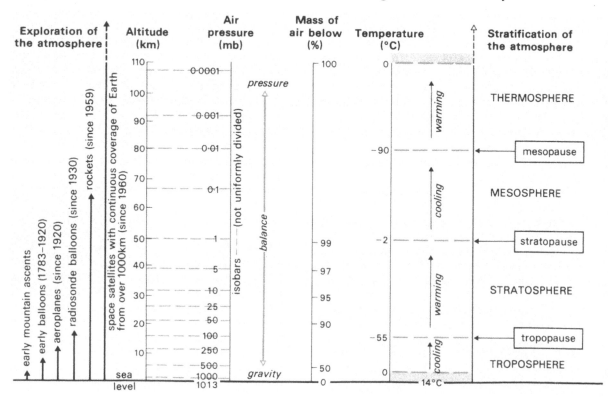

Figure 2.13 Stable layers and levels in the atmosphere

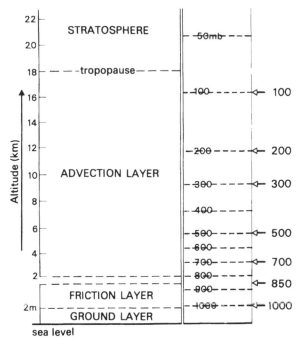

Figure 2.14 Layers and levels in a standard troposphere

and rockets for mapping purposes (see Fig. 2.2). Departures from the standard are clearly helpful in locating where pressure is higher or lower than normal. The standard situation is a frame of reference.

The equation of state
The **equation of state** relates the pressure, temperature and density of air at any place. It is easy to envisage that a pile of books remains at rest because the downward pull of gravity on any one volume must be equalled by its own *internal* force acting upwards. Likewise, the external pressures on any body at rest must be equalled by its internal pressures. Because air is not a rigid substance, however, any internal pressure must stem from the molecular activity inside it. In effect, pressure is created by the bombardment of a surface by active molecules. Temperature, on the other hand is defined by the speed at which the molecules move about. Thus, air pressure must be a function of its temperature and, of course, its density. This principle was first applied to all gases by Robert Boyle in 1659. He referred to his discovery as 'the spring of air' and showed by simple experiments in a room that pressure was *proportional* to temperature times density. In the case of convection, for example, air is warmed above a heated surface so

that it expands and a fall in pressure occurs. The converse happens above a cooled surface (see Fig. 2.6).

The thermodynamic equation
The **thermodynamic equation** explains why changes in pressure, temperature and density cause movement as, for example, in convection and baroclinic air. From the First Law of Thermodynamics (see p. 4), it can be argued that the heat energy added to a body must equal the change in its internal energy *plus the work* it does in expanding against the external pressures that initially held it at rest or static. In other words, the air is made to work. Expansion is effectively work done *by* the air, and compression would be work done *on* the air. As seen, any expansion will cause a decrease in density and a drop in pressure. This lifts the body to a level where it returns into equilibrium with its surroundings. Thus, hot air rises and cold air sinks in convectional circulations, and the same principle is used in hot air ballooning.

Another important meteorological aspect will be rai ed in the next chapter regarding **adiabatic processes** (Ch. 3.E). Because gravitation ultimately rules, the maxim 'what goes up must come down' applies. Thus, lifting imparts potential energy to the air known as its **geopotential.**

The equation of continuity
The **equation of continuity** then explains what happens when air bodies move within air. It simply states that the mass of air moving from a given space or volume must be instantly replaced by an equal one *unless* a change in density has occurred. After all, there can be no gaps left in the air! This is a clear case of the principle which says that matter is conserved. When a body of air moves its position, therefore, another takes its place and so on generating a chain of *instantaneous* motion called a **streamline.** These are localised or self-contained in the convectional overturning already described because they circulate like traffic on a busy roundabout (see Fig. 2.6).

An even more elegant development of the equation of continuity is needed in dealing with streamlines over large distances comparable to traffic on the 'open road'. Continuity is maintained in one of two ways: either the velocity remains constant and the mass varies, or the mass is constant and the velocity changes. Queues of traffic moving along a road with many lanes are a fair analogy to streamlines.

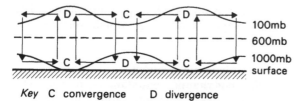

Key C convergence D divergence

Figure 2.15 Convergent and divergent air

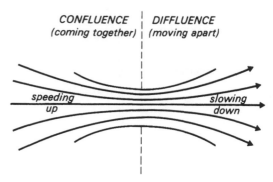

Figure 2.16 Channelled streamlines of air

When the velocity of a streamline is constant, as in free convectional overturning, a decrease in air density involves the 'thinning' or export of mass, whilst increasing density requires its 'piling' or import. The 'new' incoming air *replacing* each situation requires **convergence** of the streamlines to make up for that exported and their **divergence** to cope with imports, i.e. thinning is countered by piling-up and vice versa. Think of trying to keep the lanes of traffic moving at a constant speed when some vehicles leave or join the queues.

Figure 2.15 illustrates both and is worth comparing with Figure 2.9 earlier. Notice how the standard lower and upper isobars of 1000 mb and 100 mb are undulating because of alternate swelling and shrinking of the troposphere. This produces opposing horizontal pressure gradients above and below the 600 mb level in the case illustrated and is a clear example of thicknesses related to circulations. At the *ground surface* the convergence of streamlines is associated with cyclones and their divergence with anticyclones.

When the mass remains constant, streamlines can maintain a uniform velocity only by flowing parallel to each other, e.g. rigid lane discipline by queues of traffic or lines of people marching together. This is rarely possible over long distances and bunching or spreading is inevitable to get them 'out of lane or line'. If the lines begin to bunch together, each must speed up to avoid hold-ups. Conversely, when they spread apart, it is possible to slow down. Currents in rivers do this too on passing through narrowing channels, all other things being equal, and so does the 'crocodile' of people hurrying along a corridor if it is to avoid being trampled upon by others with similar intentions! Streamlines which come together are called a **confluence** and those which move apart give rise to a **diffluence.** Figure 2.16 illustrates the channelling of streamlines.

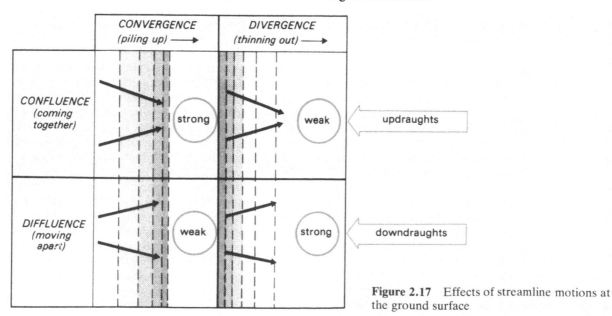

Figure 2.17 Effects of streamline motions at the ground surface

All four are brought together in a method of mapping motions called **streamline analysis.** Briefly, by using isotachs and isogons together (see Table 2.2), *rates* at which flows are converging at right angles to the streamlines are revealed; confluence reinforces convergence but counters divergence, and diffluence supports divergence and opposes convergence. This tongue twister is illustrated on Figure 2.17 with four diagrams which are essentially streamline maps applicable to ground surface situations. The combinations giving rise to strong updraughts (from the surface) and downdraughts (from aloft) are very important occurrences and are closely watched by weather forecasters. On streamline maps such places are called **singular points** and they coincide with the centres of pressure systems. When streamline maps are drawn at high altitudes rather than at the ground, of course, the resultant vertical 'draughts' must be related to the chosen level (see pp. 28-29).

The equation of motion
The **equation of motion** makes a fitting climax to this section. Hardly surprisingly, it has several applications and is founded upon the momentous work of Isaac Newton published in 1687. Three great principles are held: first, that bodies at rest or in motion will remain so unless acted upon externally; secondly, that the acceleration of a body in a given direction is the sum of all the component forces acting upon it in that direction; and thirdly, that to every action of force there is an equal and opposite one.

The first and third of **Newton's Laws** have been implicit throughout this section. Therefore, it is largely the second one that needs consideration to explain the *patterns* of motion that produce our weather.

F   Effects of the spinning earth

At this stage an important distinction must be drawn between the *driving* and *steering* forces of any motion. Their difference is that between pedalling a bicycle forwards and moving the handlebars: the latter affect the motions but do not cause them. So this section falls into two parts: resolving the steering forces and identifying the pressure patterns.

Resolving steering forces
From the familiar *horizontal* viewpoint on the map, it was seen (Fig. 2.4) that the continual

tug-of-war (or pedalling) between gravitation and convection gives rise to horizontal pressure gradients. The resultant driving force (denoted **P** on Fig. 2.4) is generated across the isobars on a pressure map at right angles from high to low. Because of the localised nature of much convection, for example, high- and low-pressure regions are cell-like systems (see Fig. 2.5) enclosed by circular isobars. Also, since the Earth rotates and its surface is curved but 'rough', three other forces must be resolved in *steering* the winds about the world. A concise description of each one is given in Table 2.3. Although many alternative names are used, we may choose the terms rotational, centripetal and frictional forces respectively. For the sake of subsequent diagrams, they are also notated **D, C** and **F.** The first is attributable to the Earth's rotation, the second results from curved paths of motion, and the third depends upon the 'roughness' of surfaces.

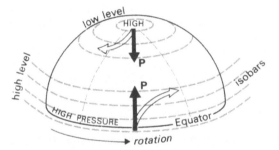

Figure 2.18   Deflection of air moving north and south in the northern hemisphere

Figure 2.18 shows the outcome of the **rotational** or **Coriolis effect** in deflecting winds. To earth-bound observers, winds appear to be steered more and more to the *right* of their initial direction of movement when viewed downwind in the northern hemisphere. On the other hand, the deflection is to the *left* in the southern hemisphere. The effect is absent over the Equator then increases in value with higher latitudes. Thus, it exerts a proportionately greater influence on steering winds in mid-latitudes and has a maximum value at the poles (see Fig. 5.1).

At high altitudes and over large uniform ocean surfaces where there is little frictional 'drag', deflections in mid-latitudes may continue until the wind blows parallel to the isobars, i.e. at right angles to the pressure gradient. Obviously, it cannot be deflected any further to go against this gradient and, thereby, comes into a state of *balance*. At this stage, the pressure gradient and

Table 2.3 Three steering forces on motions

Rotational or deflective (D) – the effect of the Earth's rotation upon its axis. Often called the **Coriolis force** after the Frenchman credited with its discovery. If a body of air seen from the ground appears to be stationary over the Equator, in fact, as seen from a space satellite it would actually be moving at the same speed as the Earth's surface rotates there. Thus, the body possesses momentum (see Table 1.1), which is best called angular momentum since it is circling Earth in the direction of rotation. Like the mass whirled around on the end of a rope, the angular momentum of the air body depends upon the Equatorial radius of the Earth. The momentum possessed must be *conserved* if the body is moved polewards by other means even though the radius of rotation gets smaller, i.e. the string is shortened during whirling. This can only happen by the body gradually increasing its speed or angular velocity from west to east with the Earth's spin. The increases will be a function of the higher latitudes.

The result is that, looking poleward from the Equator, the body appears to be increasingly deflected to the *right* of its apparent line of motion in the *northern* hemisphere. In the *southern* hemisphere the deflection is to the *left*. The same right and left rules apply to bodies moving *towards* the Equator in the northern and southern hemispheres respectively so long as the viewpoint is *downwind* (see Figs 2.1 and 2.18).

Centripetal or cycolostrophic (C) – the effect developed when a body follows a curved path of motion (such as would be caused by the progressive deflections explained above). Just as a bicycle must be leaned into a corner when going around it, so this force acts *inwards* towards the local centre of rotation or curvature. Its strength (see Table 1.1) is inversely proportional to the radius of curvature, i.e. when the bend is 'tight' with a *small* radius of curvature the inward force is *strong* and vice versa.

The result is that if the body is moving at a fixed speed, a smaller radius of curvature produces a bigger pull inwards. The faster the motion, the stronger will be the pull exerted.

Frictional (F) – the effect of the 'roughness' of the Earth's surface on the lowest air motions. Bodies moving over smooth surfaces such as the oceans or well above the friction layer (see Fig 2.14) are unaffected by this force. Rough terrain slows motions near the ground to reduce the influence of the two forces (**D** and **C**) explained above. The frictional force acts in the exact opposite direction to the motion concerned.

The result is that the horizontal pressure gradient force (**P** on Fig. 2.4) begins to exert itself more above areas with 'strong' relief. The motion is 'dragged' *towards* the centre of the low pressure crossing isobars at an angle which may reach 50°. Ekman first applied this principle to the study of ocean currents over continental shelves.

---

rotational forces are equal but effectively pulling in opposite directions. Any such balanced motion is described as a **geostrophic wind,** which simply means 'Earth-turning'.

Christoph Buys-Ballot of the Netherlands applied the above deflection rules in 1860 to determine the general position of pressure systems. **Buys-Ballot's Law** states that, when looking downwind along the line of a balanced flow in the northern hemisphere, high pressure lies to the right and low pressure to the left. The reverse sides hold in the southern hemisphere.

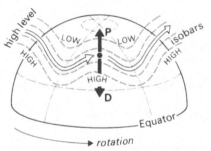

Figure 2.19 Upper westerly waves at the mid-latitude pile-up

Figure 2.19 shows high-level winds circling the Earth with its rotation in mid-latitudes. The same applies to both hemispheres. As mentioned already, air piles up in these zones so that upper-level high pressure lies to the right of these winds (above the Tropic north of the Equator) and low pressure to the left (nearer the polar regions) in accordance with Buys-Ballot's Law. It will be seen shortly that these upper-level highs and lows frequently push ridges and troughs northwards and southwards respectively to introduce wave motions aloft (Fig. 2.19). Similar waves may form in the southern hemisphere. Within such waves, the **centripetal force** described in Table 2.3 acts alternately to north and south supporting the pressure-gradient force and rotational force in turn. The resultant balance sustained by these forces is not so exact and it gives rise to the looping **gradient wind.**

Because geostrophic and gradient winds in layers stacked above each other blow in slightly different directions, whole 'beds' of air can be visualised as sliding over others with varying velocities and directions. This gives rise to degrees of **wind shear** with height. The general flow of air averaged out within a given layer is simply called the **thermal wind,** although it is not a wind in the strict sense. Careful watch is kept on the changing directions of the thermal winds aloft for reasons that will become evident in the last section (Ch. 2.G).

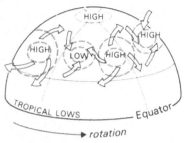

Figure 2.20 Surface pressure systems beneath the mid-latitude waves

Identifying pressure patterns

Figure 2.20 shows the more cell-like pressure systems that develop beneath the upper westerlies just described. Their spiral or gyral winds are assisted by **frictional forces** near the ground surface as outlined in Table 2.3. The 'rougher' the relief, the greater is the angle made by the resultant **surface winds** blowing across the isobars. In the northern hemisphere, the rules are: anticlockwise and inwards about a low-pressure system (or cyclone); and clockwise and outwards from a high-pressure system (anticyclone). The directions of spin are reversed for such systems in the south-

ern hemisphere but not, of course, the 'ins and outs'.

The whole set of possible patterns is summarised on Figure 2.21 with respect to the northern hemisphere once again. Each part of the diagram shows how the driving and steering forces are resolved and the ultimate balanced winds generated. With the surface systems, it is worth noting that on equal terms the winds around highs would be relatively stronger than with lows. This is rarely evident in practice because the pressure-gradient forces of lows are invariably much stronger.

In Chapter 2.C, simple convection was seen to result from heating contrasts on the ground which gave rise to different pressure systems at the surface and aloft (see Fig. 2.15). Such systems are largely of *thermal* origin, therefore, and are 'ground controlled'. In this section, however, it has been apparent that the spinning surface induces similar systems of *mechanical* origin from within the air layers themselves. It is found that both types of system favour different latitudinal zones in response to whether either thermal or mechanical processes dominate. The former frequent tropical and continental regions seasonally whereas the latter occur more in the mid-latitudes. The *relative rates* at which pressure falls from the ground surface upwards identifies each variety.

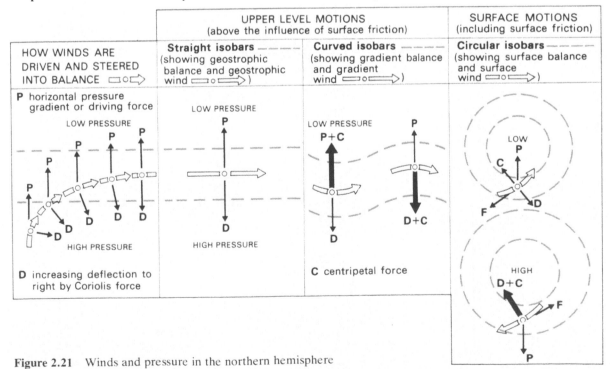

Figure 2.21   Winds and pressure in the northern hemisphere

Table 2.4 Low- and high-latitude pressure systems

At low latitude or over heated surfaces	*At high latitude or over cooled surfaces*
surface low with a warm core; pressure decreases with height slower in core relative to its cooler surrounds; weakening aloft may mean that it becomes capped with an upper high with outflowing winds (divergence)	surface high with a cold core; pressure decreases with height rapidly in core relative to its warmer surrounds; weakening aloft may mean that it becomes capped with an upper low with inflowing winds (convergence)
Examples: tropical cyclones, hurricanes and typhoons (see Ch. 5.A); convectional lows over heated lands (see Fig. 2.10)	*Examples:* polar anticyclones or inversions and stable mid-latitude highs over cold continental interiors and mountains (see Fig. 2.10 and Ch. 5.B)

Table 2.5 Mid-latitude pressure systems

At mid-latitude or beneath upper waves	
surface low with a cold core; pressure decreases with height rapidly in core relative to its warmer surrounds; strengthening aloft occurs with intensification of cyclonic circulation there; upper waves cause divergence (see Ch. 2.F)	surface high with a warm core; pressure decreases with height slower in core relative to its cooler surrounds; strengthening aloft occurs with intensification of anticyclonic circulation there; upper waves cause convergence (see Ch. 2.F)
Examples: cyclones or low pressure systems affecting the British Isles and most of northern Europe (see Chs. 4.A and 5.B)	*Examples:* anticyclones or high pressure systems affecting subtropical regions like the Azores and Mediterranean Europe (see Chs. 4.A and 5.A)

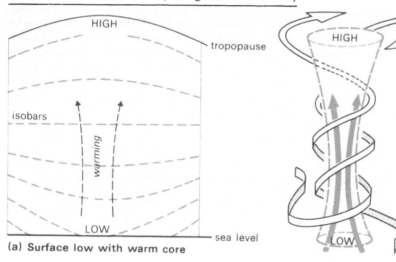

(a) Surface low with warm core

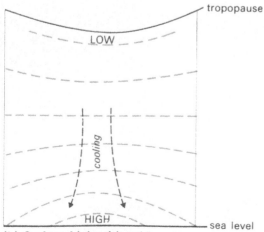

(b) Surface high with cold core

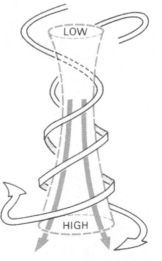

Figure 2.22 Flows in warm lows and cold highs in the northern hemisphere

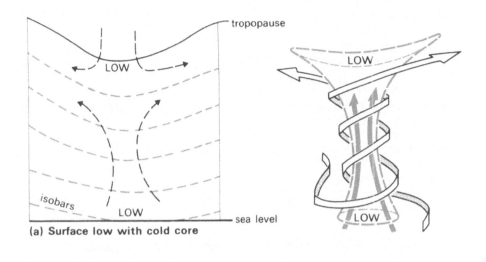

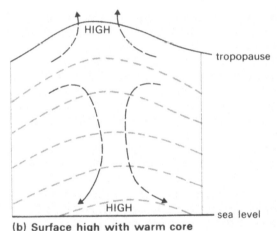

(a) Surface low with cold core

(b) Surface high with warm core

Figure 2.23 Flows in cold lows and warm highs in the northern hemisphere

Four types of surface to upper air patterns can be envisaged. These are described and exemplified in Tables 2.4 and 2.5 with illustrations of typical cases in Figures 2.22 and 2.23. Notice how the rules of motion around pressure systems in the northern hemisphere have been applied to show the motions three-dimensionally. The thermal systems above heated and cooled surfaces are clearly of a shallow convectional nature, while those associated with the upper waves have mechanical origins aloft. The latter show motions very similar to the vortices formed when hot water is stirred into a bath of cold water.

The largely mechanical systems in mid-latitudes are vortices where warm tropical and cold polar air masses meet to be stirred by the upper waves induced by the spinning Earth and forced to mix. The contract between both masses of air is called the **polar front** and the classic explanation for the mid-latitude systems was formerly referred to as the **polar front theory** because it relied more upon general revolving and actual mixing by frontal activity rather than upper waves. With the wealth of information that now exists about the upper air waves, however, this theory has been found to be too general. A better explanation emerges when the entire atmospheric circulation is studied from the standpoint of distributing heat and mass energy uniformly over what amounts to an irregular moving surface.

G Global motions

All the threads of evidence, whether thermal or mechanical, are now brought together to account for the **general circulation** of the atmosphere.

The low- and high-latitude systems are readily explained in terms of the convectional overturning of the tropical Hadley cells and polar cells respectively (see Fig. 2.9). However, the revolving mid-latitude systems pose greater problems and clearly do not fit a simple advective model as proposed by Ferrel. Although their intensification aloft indicates origins within the upper air streams near the roof of the troposphere, we shall see that the ground also plays a part both thermally and dynamically. Therefore, this section breaks down into three parts: the principles governing revolving pressure systems; wave motions in the upper air streams, and the effects of land masses and their relief.

Revolving pressure systems
Localised and spinning columns of air moving over the rotating Earth are governed by the same rules that influence giant eddies or vortices in fluids. The similarity between the flows on Figure 2.23 in particular with those of water sucked down the plug hole of a bath is marked. The principles of spinning fluids, aptly called **vorticity**, were developed in 1845. Almost a century later they were used by meteorologists led by Carl Gustav Rossby to explain the upper waves and surface-pressure systems in mid-latitudes.

Vorticity has two forms: that possessed *locally* by the spinning column on its own account and that imparted *globally* by the rotation of the Earth. The former is usually called the **relative vorticity,** and the latter may be thought of as the **global vorticity.** Added together they comprise the **absolute vorticity** of the spinning column. It can be shown that the absolute vorticity is effectively conserved in large spinning air bodies with a given vertical pressure difference between the top and bottom of the column concerned. In other words, the rate of change of local relative vorticity *plus* global vorticity in a given column must always be zero, as we shall now see.

Figure 2.24 shows three similar spinning columns at different latitudes: over the Equator (A), at the Pole (B) and in mid-latitude (C). It can be seen that column A is at right angles to the rotational axis of the Earth, B is actually aligned with it and C is at an angle which is a function of its latitude. Clearly, A cannot have any element of global vorticity given by the Earth's rotation, B will have most, and C will be intermediate depending upon its latitude. Put simply. if you are standing at the Equator and wish to spin around 'on the

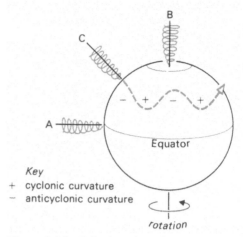

Key
+ cyclonic curvature
− anticyclonic curvature

rotation

Figure 2.24 Spinning air and wave motions

spot', you will get no help from the rotating Earth. Conversely, at the Pole, you *are* already being spun around whether you wish to be or not!

Now, if a spinning column has to move into higher latitudes, its global vorticity must increase at the expense of its relative vorticity because of their overall conservation. When moving into lower latitudes the reverse occurs. Thus we have a see-saw effect between both forms of vorticity. Instead of wandering at random, therefore, this sets up a regular wave motion. The *tendency* to wave will be at a maximum in mid-latitudes where the two are equally poised. All that is required is something to start the waving off, and this will be discussed shortly. Notice in Figure 2.24 that the convention in the northern hemisphere is to call equatorward loops **positive** or cyclonic and poleward loops **negative** or anticyclonic.

Wave motions in the upper air streams
Figure 2.25 shows the principal upper- and surface-pressure systems associated with the mid-latitude or **Rossby waves** aloft (also, see Fig. 2.23). It is apparent that cyclones develop below waves going north and anticyclones occur beneath waves going south. Cyclones have cold cores sloping upwards towards the north west ahead of the upper trough and anticyclones have warm cores sloping upwards to the south west ahead of the upper ridge. This alternate low and high pattern is the distinctive feature of British weather (see Ch. 4.A).

At the surface, lows in mid-latitudes are regions of maximum relative vorticity with flows which converge and ascend. This must be balanced by

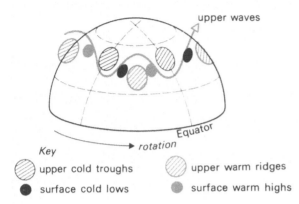

Key
upper cold troughs   upper warm ridges
surface cold lows   surface warm highs

Figure 2.25 Upper waves and mid-latitude pressure systems

high-level divergence along the northbound upper air streams (Fig. 2.23a). The converse applies to highs at the surface which spin more slowly and contain descending warm cores which diverge at ground level. Above them, the southbound upper air streams are characterised by convergence (Fig. 2.23b). In simple terms, the lows are being *pulled up* and the highs being *pushed down*.

On a northbound run, decreases in relative vorticity take place compared with increasing global vorticity in the upper wave (see Fig. 2.24). This 'take over' by the rotating Earth effectively 'opens up' the upper flows. Thus, horizontal divergence or thinning out aloft occurs to generate the necessary pull upwards. Along a southbound run, by contrast, increases in relative vorticity are favoured against global vorticity. The decline in Earth turning 'shuts up' the upper flows. This causes horizontal convergence and piling up aloft to give the necessary push downwards. Another consequence of these controls (considered further in Ch. 5.A) is that pressure systems nearer the Equator become non-revolving and virtually stationary. At higher latitudes we are accustomed to them revolving and moving rapidly with the Earth's rotation.

Sometimes either confluent or diffluent characteristics will be imposed in addition upon the streamlines within the upper waves (see Fig. 2.16). These conditions result from relatively greater thermal swelling and shrinking of the main upper ridges and troughs causing horizontal pinching together in the case of a confluence and spreading apart in a diffluence (see Fig. 2.16). Whichever combination takes place (as in Fig. 2.17), the consequent strengthening or weakening of vertical motions will play a big part in the vigour of the associated surface systems. Such behaviour justifies the close watch kept by meteorologists and forecasters, especially on events in the upper air streams over mid-latitudes, e.g. the changing directions of thermal winds are tell-tales of warm air pushing polewards or cold air moving equatorwards.

Often, the effects of upper confluences are enough to generate high-level winds exceeding 30 metres per second (or 108 km h^{-1}). These strong meandering flows within the upper waves are called **jet streams.** The weather systems resulting from jet streams are outlined in Chapter 5.B with reference to the **index cycle** (see Figs 5.5 and 5.6).

Northbound divergent streams exert such strong controls over the formation and subsequent location of cold-core lows that they are referred to as being **cyclogenic,** meaning simply 'cyclone-generating'. Likewise, southbound convergent streams appear to be a major reason for warm-core highs and may be regarded as **anticyclogenic.**

In the foregoing model, the principles and solutions are largely mathematical. It would appear that the long-wave upper westerlies or Rossby waves could form *anywhere* within the mid-latitude pile-up from subtropical to polar latitudes. Although actual observations support this view to some extent, the search for mechanisms to *start* the waving still continues. Most attention has centred around the apparent fact that preferred positions exist which anchor the waves for long periods. During such times, they become effectively **stationary** or standing waves, being well-defined and meandering channels for rivers of upper air to circle the Earth. At other times greater flexibility is evident and the waves become **progressive.**

Effects of land masses and relief
It is with the stationary or progressive behaviour of the upper waves that the distribution of land masses and their major relief features appear to exert their strongest influence upon the general circulation. Two broad features seem to be critical: the general distribution of land, sea and ice; and the alignment of the high mountain ranges. Clearly, then we should expect differences between the comparable circulations of the northern and southern hemispheres. Furthermore, these differences should reveal what roles are played by both sets of features.

Figures 2.26 and 2.27 are viewpoints of both hemispheres from above their respective poles; latitudinal zones are concentric and meridians

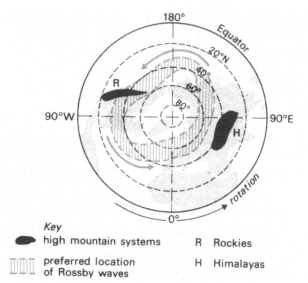

Figure 2.26 Course of Rossby waves in the northern hemisphere

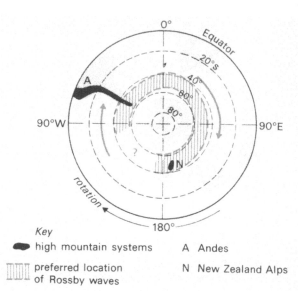

Figure 2.27 Course of Rossby waves in the southern hemisphere

radial like a spider's web. On such maps, Rossby waves appear uninterrupted as vast zonal rivers circling the poles. This is sometimes called the **circumpolar vortex.**

Compared with the almost true zonal ring in the southern hemisphere, the northern vortex is obviously distorted. Two cold polar troughs push southwards across the eastern margins of the great continental land masses at high latitudes in the northern hemisphere. Corresponding warm ridges thrusting northwards seem to be linked with the high mountain systems shown, particularly the Rockies in North America. Except for a suspicion that the southernmost tip of the Andes and, maybe, the New Zealand Alps might produce similar effects, the circuit in the southern hemisphere is largely above oceans and is, therefore, uninterrupted.

The cold troughs over the north-eastern margins of Canada and Asia suggest origins associated with strong *thermal* contrasts where continental coastlines and ice masses meet furthest south. If so, they should be more prevalent during the onset of winter snows and spring thaws as the air layers above snow surfaces are chilled more than those over 'bare ground'; consequently, bigger regional temperature *differences* exist either when the first snows arrive and cooling occurs or as they melt and warming up takes place. Such circumstances favour the development of jet streams.

Mountain barriers over 2 km high must present

strong mechanical or *dynamic* controls as the upper waves soar over and around them. Increased friction (see Fig. 2.21) occurs with horizontal divergence as the air climbs the peaks on the windward slopes. Together, these turn the upper winds to the north and generate cyclones below. On the leeward slopes, decreasing frictional effects and horizontal convergence as the air dives across the following plains will turn the upper winds to the south and generate anticyclones. The see-saw effects of vorticity assist these processes.

Figure 2.1 at the beginning of this chapter anticipated this situation and important consequences will be discussed in later chapters. Suffice it here to point out that some researchers consider that the cold troughs at the eastern margins of continents and the warm ridges above mountain barriers can get 'into phase' with respect to the mid-latitude vorticity see-saw. When this happens the upper waves become doubly anchored and stationary. This causes prolonged spells of persistent weather. The contrasting British summers of 1976 with a drought and 1978 with much cloud and rain are probably instances of such activity. By contrast, other researchers claim that this behaviour can be explained mathematically without recourse to the geography of the surface. They do this with even more refined developments of Rossby's mathematical model, and the debate will undoubtedly continue.

Short of removing the Rocky Mountains and

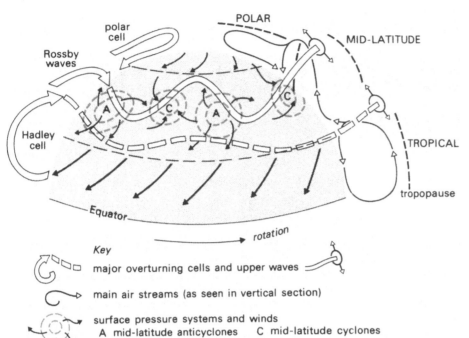

polar
cell

POLAR

MID-LATITUDE

Rossby
waves

C

A

C

A

C

Hadley
cell

TROPICAL

tropopause

Equator

rotation

Key

major overturning cells and upper waves

main air streams (as seen in vertical section)

surface pressure systems and winds
A mid-latitude anticyclones   C mid-latitude cyclones

Figure 2.28 The general circulation of the atmosphere

continents, it is difficult to be sure about their connection with the general circulation. However, it would be a shame if, towards the close of a long and largely theoretical chapter that has at last 'come to earth', we take off yet again into the realms of theory!

Figure 2.28 shows the broad features of the **general circulation** that are currently envisaged. It shows the overall global motions resulting from the processes dealt with in this chapter and is largely attributable to the work of Palmen (1951). Notice how the tropopause has three steps towards the poles representing the varying thicknesses of the troposphere in the tropical, mid-latitude and polar regions. Breaks in the steps are associated with the strong polar front jetstream and weaker subtropical jetstream shown as Rossby waves. Also, the strongest up and down mixing clearly takes place in the mid-latitude systems beneath these waves, particularly along the polar front (which is shown as an inclined broken line on the right side of the model). Remember that this is the 'battle front' between warm tropical air moving northwards and cold polar air pushing southwards (see Chs 4.A and 5.B). Something has 'to give' where they meet.

At the end of Chapter 1 it was noted that the atmospheric machine was merely ticking over and would soon stop without solar energy to drive it. Moreover, it can now be added that the rotating and uneven surface stirs or steers the motions very sluggishly for its amount of contact with the atmosphere. If started from rest, it would be over a month before the general circulation was in full working order.

It is found to 'tick over' about every 14 days. In other words, it takes about a fortnight for air to circle the Earth. The great scope for studying repeating cycles must be the envy of geomorphologists struggling with much longer time scales. Small variations in the roles played each time round are regarded as disturbances or **perturbations** to the basic model which change the *weather*. Indeed, the whole system is very quick to adjust and accommodate changes, despite its complexity. On the other hand, there also appear to be permanent changes in the roles which lead to longer-term disruptions or **alterations** thought of as *climatic* change. The following chapters pursue these lines of enquiry by concentrating upon the surface processes associated with everyday weather and concluding with the possibility of changing climates.

Chapter 3

Moisture and Heat in the Atmosphere

The word 'weather' stems from the sense in which the atmosphere is being 'blown about'. So, the questions must be what, apart from air, is being blown about and how this happens.

The purpose of this chapter is to seek out and explain in more detail the processes by which *mass* and *heat* energy are moved about in the air. Water comprises an important part of the mass to be moved and temperature defines the condition or *state* of the heat energy involved. In the air, most water is in its highest energy state as a vapour. Its intermediate condition as a liquid is common enough, however, and the lowest one as frozen snow or ice is not unusual. All three conditions frequently occur in the weather of the British Isles.

It is hardly surprising that moisture and heat together are the most important factors regulating the condition of all living things too. Thus, it is appropriate to dwell upon some of their special environmental roles nearer the ground rather than their broad global distributions. Nevertheless, even here we shall be dealing with flows or *transfers* of energy from one object and place to others.

A Transfers of energy

It has been seen already that energy must always flow from its highest to lowest condition. This is a gradient which, like any slope, has magnitude and direction along which the transfers of mass and heat occur. In this section, the processes concerned are introduced with particular reference to their effects upon Man as a high-energy body surrounded by a lower-energy atmosphere. A discussion of the six forms of energy transfer in the atmosphere at large follows this.

Flows of water and warmth
Our personal regulation of water and warmth is a good way to appreciate the controls on flows of energy between bodies. The water content of almost all living organisms exceeds 70% of their total body weight. It is their chief source of mass and also the main means by which sensible heat energy is stored and transferred to and from the body. The average adult person, for example, contains about 47 litres of water amounting to just under three quarters of the body's weight. Normal body temperature is close to 37°C.

Different body functions control the internal transfers of water and heat. These respond in turn to external mechanisms stimulated by behaviour and *contact* with the atmosphere. This contact or **interface** is where an equilibrium between the body and surrounding air is most sensitive because the maximum energy gradient occurs there. All organisms are sensitive in this respect, each species having become adapted to particular environmental zones or niches which suit its behaviour. The range of environments occupied reflects the organism's tolerance.

Man is a more tolerant organism than most, owing to his adaptable behaviour. Yet, a person's water content must remain within 30% of the necessary 47 litres and body temperature to within 10% of 37°C.

The reader, therefore, who is probably at rest in a room with adequate amounts of moisture and warmth, provides an energy gradient for transfers of both mass and heat from the person to the surrounding air. Each interface along the gradient is the contact or surface at which the transfers occur. In order, these would be the skin surface,

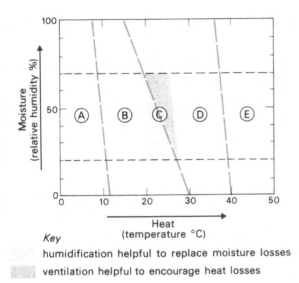

Key

humidification helpful to replace moisture losses

ventilation helpful to encourage heat losses

Figure 3.1   Moisture, heat and comfort

the clothes worn and the walls of the room and building to the atmosphere out-of-doors.

The transfers along this gradient are being detected by the sensations of comfort experienced, which are hopefully pleasant. For a normal healthy person, discomfort or unpleasantness is experienced in two senses: first, if either or both of the flows considerably speed up and secondly, should one or both drastically slow down. The former is worst in rooms which are too dry and cold and the latter when they are very damp and hot. These give rise to chilly and muggy or sultry sensations respectively. Other variations such as too dry and very hot make one feel sleepy, and damp and cold air create disagreeable clammy feelings.

Excessive outflows of water and warmth from the body can be lethal in certain harsh climates. Exposure to such weather can lead to irreversible dehydration and hypothermia respectively; the former is a killer in deserts and the latter a cold-climate and mountain hazard. Restricted outflows are equally fatal, of course, as in the case of a fluid build up (oedema) and a high fever called hyper-pyrexia. These are caused by medical disorders, however, and so are pathological rather than environmental. Happily, we are dealing with the latter here.

Amounts of moisture held by an air body can be expressed in percentage terms as **relative humidity,** values (see Ch. 3.C). On the other hand, levels of sensible heat are measured by degrees of **temp-**

erature (see Ch. 3.D). Significantly, of course, temperature standards and scales are based upon the freezing, boiling and vaporising points of water under controlled conditions. Relative humidity, too, is determined by taking air temperatures. These links reflect the mutual relationships between water and warmth.

Figure 3.1 shows how humidity and temperature relate to human comfort levels. The box labelled C is a simplified zone within which tolerable room atmospheres normally fall, or outdoor ones for that matter. It slopes slightly because both colder and warmer air feels more pleasant when drier. Notice that ventilation by light draughts and winds up to about 5 m s^{-1} (or 15 km h^{-1}) improves sultry conditions, and humidification helps to prevent wilting or even dehydration in drier air.

The other zones labelled on Figure 3.1 indicate less tolerable or more uncomfortable environments. These must be combated in appropriate ways; by continual heating in the case of A, additional warmth and humidity controls in B, heat extraction and air conditioning in D, and continual refrigeration in rare situations like E where temperatures exceed body heat. Out-of-doors each one needs suitable clothes, foods and levels of activity to regulate the energy transfers.

Forms of energy transfer

Figure 3.2 shows six forms of energy transfer: radiation; conduction; convection; evaporation; condensation, and precipitation. Figure 3.2 illustrates how these processes are linked together with convection playing the main role in cycling both moisture and heat energy in the air. Details of each process in the upward and downward transfers are highlighted in Tables 3.1 and 3.2 respectively. They repay careful study.

Radiation and conduction not only transfer heat energy but they work closely together at the ground surface and are chiefly responsible for the unequal warming that stimulates convection (see Fig. 2.6). When this is vigorous, evaporation off surfaces speeds up to transfer both mass and heat energy upwards. Once aloft, the weakening of convection allows condensation and precipitation to return the mass and remaining heat energy earthwards. It will be seen that the six taken together form the basis of both the water and heat cycles in the atmosphere.

We may return to the human body again to emphasise the significance of the forms of energy transfer. Assuming a constant relative humidity of

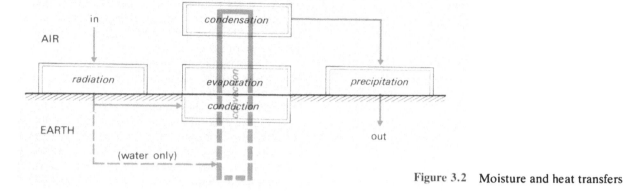

Figure 3.2   Moisture and heat transfers

Table 3.1   Four main transfers from the ground to the air

Radiation. This is energy transferred by electromagnetic waves (see p. 2). It affects heat alone because exchanges occur through space rather than by touching objects. Flow rates are governed by the temperature and texture of the emitting surface. Perfect or ideal surfaces are called **black bodies.** Exchanges can only take place between bodies which are *in sight* of each other, e.g. radiant warmth from a fire is only felt when in sight of it or the chilling effect near a window because of personal radiation directly towards it. In both cases, the effects can be reduced by moving out of the line of sight or by placing something between like a screen or curtain.

Conduction. This requires objects to be in contact with each other so that the temperature condition can be passed along from the region with most heat to the one with least. It affects heat energy alone and applies mainly to solids in which no transfers of mass can occur; therefore, it is important in rocks and soils but not in the air. Heat energy is conducted away from the body by the chair being sat upon, the wall leaned against and even the floor one is standing upon *unless* any of these surfaces have temperatures above 37°C.

Convection. This mechanism has been introduced already as the chief energy transfer process in the atmosphere (see p. 15). It only works *in contact* with an unequally heated surface like the ground and involves both vertical and horizontal motions, the latter being called advection. Bodies of air are moved, so mass transfers occur with heat energy literally carried by the molecules on the move. Free or forced convection can take place. Both processes stimulate others, notably evaporation. The higher wind speeds with forced convection increase chill factors by accelerating mass and heat losses. Clothes are simply means of 'keeping out' convection.

Evaporation. Here water transfers from a liquid to a vapour state, extracting both mass and its 'hidden' stores of latent heat energy together (see p. 39). Flow rates are governed by temperatures, amounts of moisture already in the air at the evaporating surface and the vigour of the convection process. Respiration (breathing) and perspiration (sweating) by animals and transpiration by plants are special mechanisms linked with evaporation because they remove moisture and heat from the body by vaporisation into the air. Evaporative cooling is a vital heat regulator in all living things on land. Strong flows give rise to rapid drying and chilling, both of which can be dangerous in excess.

Table 3.2   Two main transfers from the air to the ground

Condensation. This is when water vapour turns into liquid droplets and releases its 'hidden' stores of latent heat energy together (see p. 39). The process is controlled by temperatures, amounts of moisture, salt and dust present and the vigour of convection. Humid air mixed suddenly with cold air produces a fog or, if it contacts a chilled surface, forms a dew. The former is seen when breathing out warm moist air on a cold day and the latter when window panes 'mist up'.

Precipitation. This is when liquid water droplets or frozen ice particles fall under the influence of gravity (see p. 53). The process is controlled by temperatures and the growth of water droplets and crystals sufficient to overcome upward convection currents. Rainfall is the chief form of precipitation but hail, sleet and snow occur in colder conditions.

about 45%, it is possible to determine the proportions of moisture and heat energy lost from a body owing to radiation and convection as against evaporation. At a surrounding or **ambient** air temperature of 15°C nearly 90% is lost through radiation and convection and the remaining 10% by evaporation. At 26°C the proportions are roughly equal whereas at 37°C over 99% must be lost by evaporation. This, of course, explains why profuse sweating accompanies a fever.

Some of Man's instinctive behavioural responses to uncomfortable conditions are of interest, in conclusion. 'Goose flesh' appears to be a rather futile way of trying to reduce convectional effects close to the skin in cold weather, and curling up in bed decreases the surface area exposed relative to the body's mass. Conversely, in hot conditions, involuntary sweating increases evaporative losses (see p. 39) and limp postures help to expose the largest possible surface area relative to mass. An awareness of such responses will help in understanding the key atmospheric processes at work.

B   The nature of air

Like all substances, air is made-up of the fundamental elements and, so, the tiny 'worlds' of atoms. Unlike the reactive compounds forming minerals and rocks, however, air is largely a highly active *mixture* of individual or isolated gas molecules. Yet, it is neither wholly clean nor dry. The presence of solids and liquids are as vital as the gases in governing atmospheric processes. The role of water is sufficient to warrant highlighting in the next section whereas here we may review the solid and gas ingredients. Appropriately, these suggest how the whole atmosphere evolved to acquire its particular composition today.

The origin of air
Elements may be grouped and graded by the masses, sizes and shapes of individual atoms. Of the 92 natural elements known, those mixed in air come from the top ten lightest and largest atoms with the simplest structures or configurations. The dominant gas constituents are nitrogen and oxygen; but, this has not always been so as the presence of other heavier rare gases indicates, i.e. argon, neon, krypton and xenon.

Atoms are the smallest units capable of reacting chemically. Each has a central nucleus holding positive electrical charges surrounded by a number of electrons carrying negative charges. When these are equally balanced the system is neutral. The greater bulk of the former determines the *mass* of the atom. Electrons describe wave-like orbits making concentric shells around the nucleus in different planes. The radius to the outer electron shell or orbital gives the *size* of the unit. Its configuration or *shape* is recognised by the number of shells and distribution of electrons among them. Inner shells may have 2 electrons, the second up to 8, the third up to 18 and each succeeding one as many as 32. Therefore, inner shells have lower energy levels than outer ones. Since hydrogen heads the list with a single electron orbital, it is the standard to which all may be related. Although beyond the scope of this book, it is significant that the Sun possesses an abundance of hydrogen atoms which fuse to produce solar energy (see p. 42).

No outer shell can hold more than 8 electrons for when such a stable octet is attained, further increases of electrons start new outer shells while filling vacancies. Elements without 8 outer electrons and so with unfilled spaces readily react with others. Those which have 8 and are full-up, on the other hand, are unreactive or *inert* like the rare gases named earlier in this section. To become stable or neutral, elements may lose outer electrons, which empties the shell, or gain more so that it fills up. This behaviour enables a broad distinction to be made between metals with positive charges (**cations**)and non-metals with negative charges (**anions**) respectively. The amounts of charge available in such ions are called their valencies; a value measuring the combining or replacing 'strength' compared with hydrogen. Reference to a **periodic table** used by chemists will show that most rock-forming minerals comprise both ions, while the stable and variable constituents of the atmosphere are mainly anions.

Ions with equal and opposite charges attract each other forming closely packed molecules. Such *ionic* bonds, however, do not hold the outer electrons tightly and, so, their molecules can easily react *chemically* with others, as is evident in the weathering of rocks. By contrast, the preponderance of similarly charged ions in air means that neutral molecules can only form in the atmosphere if outer electrons are shared by two atoms. Such paired or diatomic *covalent* bonds are the strongest of all chemical ties and, therefore, give rise to independent molecules whose *physical* behaviour predominates.

One physical characteristic is the *mobility* of

covalent gas molecules. This is governed by heat energy and their density so that excitation and collisions increase in hotter and 'thicker' air. On colliding, some molecules speed up to exceed the mean velocity of all present while others slow down. Since this activity is random, the proportions above and below the mean can be determined statistically. More collisions increase the *probability* that some will reach high enough speeds in the right direction to 'bubble off' into space. To escape Earth's gravity, upward velocities must exceed 11 km s^{-1}. These risks become higher nearer the outer limits of the atmosphere. Here continuous exposure to the ionising effects of solar and cosmic radiations is greater. Processes within the roof of the atmosphere or **ionosphere** may be enlikened to those of a protective skin. Past variations in the protection it affords have been crucial to the atmosphere's evolution.

The first 'atmosphere' enveloping Earth did not last long because high temperatures drove off its gases into space. It is thought that it comprised lethal hydrogen compounds with bromine (Br), chlorine (Cl), fluorine (F), iodine (I), sulphur (S) and nitrogen (N). The last one would have produced ammonia (NH_3). The second atmosphere, therefore, probably derived from volcanic gases exhaled from the interior as the Earth cooled. Comparisons with volcanic exhalations today indicate that the principal constituents would have been carbon dioxide (CO_2), water vapour (H_2O) and covalent nitrogen (N_2). No dissociated or *free* oxygen (O_2) existed at this primaeval stage.

The problem of how oxygen was freed remains debatable. Of the alternative theories, the most accepted proposes the **photosynthesis** of carbon dioxide by plants as the main process involved. This presupposes cooling below 100°C to begin condensation and fill the oceans. Excesses of CO_2 could then be dissolved into carbonates such as limestones and dolomites. Initially, photosynthesis would have been confined to bacteria in the upper levels of the oceans before plants colonised the land. Fossil and rock evidence bears out such a sequence up to and during Palaeozoic times to culminate in the vast coral seas and luxuriant forests of Carboniferous times about 350 million years ago.

We must conclude that the origin of the gas mixture comprising air was the product of living organisms and not the other way around. Furthermore, it is important to realise that the production and consumption of both gases and solids maintaining the present atmosphere must balance to retain its state of equilibrium. Until Man introduced large-scale industrial processes, this balance was maintained by the products of volcanoes and radioactive rocks being absorbed by organisms, particularly plants, then recycled through the land and ocean systems.

Under the prevailing temperature and density conditions in the ionosphere especially, helium (He) is the only gas continuously produced (by radioactivity) which is light enough to be 'exhausted' into space. The disposal of waste gases and solids into the air along with the removal of large tracts of vegetation (see p. 37) poses long-term problems respective to the vital production–consumption balance of its key constituents.

The constituents of air

Air consists largely of negatively charged covalent molecules of nitrogen (N_2) and oxygen (O_2) as shown in the left-hand pie chart on Figure 3.3. Being closely similar chemically, their proportions would be much the same if taken by volume rather than weight. By contrast, the right-hand pie chart on Figure 3.3 showing the main constituents of the Earth's crust would have 93% as oxygen if redrawn according to volume. Thus, oxygen is the most abundant element of the Earth–atmosphere system. Even the remaining 1% of the air is dominated by recombinations of oxygen.

All the constituents of the air have particular functions. Again, leaving a consideration of water to the next section, we may deal with the gases and solids here to assess their roles in atmospheric

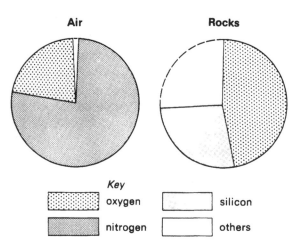

Air   **Rocks**

Key
oxygen   silicon
nitrogen   others

Figure 3.3 Main constituents of the air and Earth's crust by weight

processes. For the moment, therefore, we concentrate upon *dry* air.

Although it is a freely available and an important constituent of living matter in proteins, nitrogen is not utilised directly. Micro-organisms in the soil are necessary to *fix* it in compounds which are then cycled among all plants and animals. Three hydrogen ions readily combine with an available nitrogen ion to form the essential compound ammonia (NH_3). Also, when combined with other cations, salts called nitrates (with NO_3) and nitrates (with NO_2) are made. Because these processes occur in the ground, however, they cannot be claimed as strictly 'atmospheric'; yet it would be foolish to dismiss soil air as unrelated to that overhead, of course.

Apart from free oxygen being essential to breathing, two other forms deserve mention now: the vital triple molecule called **ozone** (O_3) and **carbon dioxide** (CO_2) play key roles in the upper and lower air respectively. Ozone is produced and concentrated into thin layers within the stratosphere. Here, solar radiation breaks ordinary oxygen into separate atoms which then collide and recombine into unstable O_3 molecules. Although continuously created and destroyed, the layers of ozone act like outer shields defending Earth against dangerous solar radiation (see Ch. 3.D).

Carbon dioxide is also continually produced and separated by animals and plants. We shall see later (Ch. 3.D) that its concentrations are the chief 'setter' of air temperatures both locally and globally. Some is assimilated in the seas and the present average in the atmosphere of about 315 parts of CO_2 per million of air appears to be rising (see p. 82). Industrial activity and, maybe deforestation are thought to be responsible because burning fuels increases the output of CO_2 while its intake by plants for recycling into fixed carbon and free oxygen is reduced.

Gas molecules may be ionised when an external energy source is sufficient to 'dislodge' an outer electron (see p. 35). This leaves the molecule with a positively charged residue and the ejected electron with a negative charge. Being now more reactive, both quickly attach themselves to uncharged molecules to form ion pairs. Different types may be distinguished by their mobility; smaller ones are 'fast' and larger ones are 'slow'. A natural balance exists between their production and destruction with fast ones prevalent over the oceans and slow ones more so above the continents.

Most minute particles of solid matter in the air, such as salts from seawater, airborne dusts and specks of manmade wastes, are also very reactive. They are called **Aitken nuclei.** Ionised gas molecules and Aitken nuclei attract each other. Such 'captures' produce even larger and slower molecular clusters which may contain harmful concentrations of acids in badly polluted urban air. These do not disperse easily. Exhaust gases and solids from cars and factories are examples which are carefully monitored.

In fine weather, an electrical current of some 1000 amps worldwide flows earthward. Ionised gases and Aitken nuclei are important conductors of this electricity from the ionosphere to Earth. Disturbed weather disrupts the flow, and its field may even be temporarily reversed in surface dust and sand storms. Static electricity is generated in these by particles rubbing together. Thunder clouds help to replenish the charge aloft because positive ions are separated out and lifted within the frozen cloud tops (see Ch. 3.C, 3.E and 4.D). Accumulated electricity is suddenly discharged and earthed in strokes of lightning.

It is clear that the constituents of air are being constantly recycled and distributed in time and space, but it is also apparent that the delicate balance between their breakdown and renewal must be altered by human activity. As will be seen, city air with added fumes laden with nuclei gives rise to weather different from that in the surrounding countryside. Nuclei mix in and float about the air in varying concentrations. In one breath you could inhale up to a thousand particles, and considerably more in heavily polluted atmospheres.

The composition of air
When mixed together, the blend of molecular processes characterising the individual constituents described helps in a broader way to generate the weather. We can conclude this section by turning to their effects upon air pressure as the primary element driving atmospheric motions (see Ch. 2.C).

Each constituent gas in the air exerts its own **partial pressure** independent of all others. When added together, the sum of all partial pressures amounts to the *total* pressure of the air at the altitude concerned. Any one contribution is a function of the molecular mass of the particular gas and its percentage of the whole mixture. At mean sea level, therefore, with a total atmospheric pressure averaging just over 1013 mb, nitrogen contributes

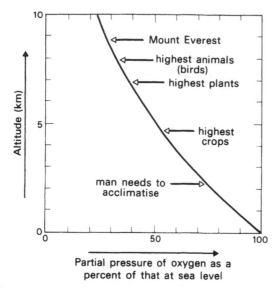

Figure 3.4   Oxygen and altitude

760 mb and oxygen 240 mb, rounding off at 1000 mb. The remaining 13 mb represents the contribution by all other gases present.

So dominant are the proportions of nitrogen and oxygen that they remain in this ratio with altitude as the overall pressure falls. Figure 3.4 gives a curve which shows how oxygen alone decreases with altitude. One for nitrogen would be similar. By comparison with its partial pressure at sea level, oxygen drops to about three quarters at just above 2000 m, to nearly half at 5000 m and to barely a quarter at 10 000 m. Notice how the highest known altitudes at which crops are cultivated in the Himalayas are at levels where oxygen is reduced by half its sea level pressure. The same applies to nitrogen. Thus, air becomes thinner or *rarefied* with height.

Most of the remaining 13 mb of air pressure to make up at sea level consists of water vapour. This is an addition to the dry air mixture, and amounts vary from almost nothing in the driest and coldest environments to 5% of air masses above the warmest tropical seas. Although tiny, such amounts are enough to make the air behave more like a fluid.

C   Moisture in the air

With its simple chemical formula and small absolute contribution to the total mass of air, water may seem to be fairly unremarkable. Nothing is further from the truth, however, because water possesses strikingly different properties than other substances of similar simplicity. These boost moisture into the position of being the second chief weather element after air pressure. Since well over 99% of all the mass of water on Earth occurs in the oceans, ice caps and ground, much less than 1% is being cycled in the atmosphere at any time. Moisture in the air, therefore, is a good case of a little achieving a lot. In this section the processes highlighting the importance of water in the air are explained.

The states of water

Figure 3.5 shows that a single H_2O molecule has a biased arrangement of electrons. Two unused pairs of outer electrons give a negative side to the exposed region of the oxygen. On the other side, the shared pairs with the bonded hydrogens present a positive region. Thus, although stable or neutral, water molecules orientate in an electrical field with their negative regions facing the source of energy. Such biased molecules are called dipoles. Likewise, the appropriate side or pole will be loosely attracted to the surface of an available ion. This is why many salts readily collect shells of water which isolate them in solutions and the air. Most raindrops, for example, are really shells of water around such salts or Aitken nuclei (see p. 37). Molecules of water vapour seem to cluster around suitable particles rather than disperse freely. Such particles are termed **hygroscopic** because of their ability to attract water vapour.

The ability to bond to another substance is called *adhesion* and that which describes bonding among molecules of the same substance is referred to as *cohesion*. Water is clearly very adhesive with respect to hygroscopic nuclei but, as temperatures rise, the more mobile molecules do not cohere well. This helps to explain why puddles evaporate

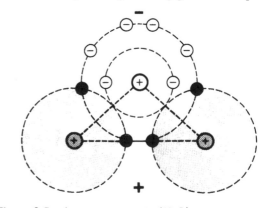

Figure 3.5   A water molecule (H_2O)

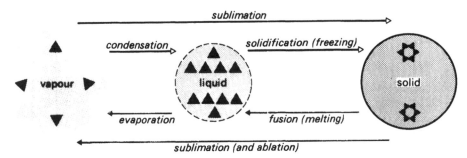

Figure 3.6 The changing states of water

quickly while the surrounding muds remain wet. As temperatures fall, the bonding rapidly gets stronger. With freezing the molecules cohere into regular tetrahedra. Each resultant ice crystal presents a positive outer surface which attracts more water molecules. Thus, ice crystals grow in layers by a process known as **accretion.** These in turn knit into the efficient hexagon structures typical of the snow flake. The variety of snow and ice conditions encountered in mountains reflect different ways in which the crystals pack together.

The melting and boiling points of water of 0°C and 100°C respectively are many times *higher* than for similar substances with low molecular weights. Otherwise, all moisture would vaporise at normal atmospheric temperatures and there would be no liquid water and no life on Earth. Figure 3.6 shows the changes of state that water readily undergoes on warming and cooling (compare it with Tables 3.1 and 3.2).

If we started heating a block of ice, it would gradually melt into liquid water which would then come to the boil and eventually evaporate into the air. In effect, the most mobile molecules escape the attractive forces of their neighbours until all become individual gas molecules. This is why the less mobile residual molecules experience a cooling effect. The chilling felt when wet clothes or skin dry out is a good instance of evaporative cooling. Two changes of state have occurred and the energy used to achieve each one has been stored as 'hidden' or **latent heat.** At the melting and evaporation stages, this is locked up as potential energy because decreases in cohesion allow greater freedom for movement or mobility of molecules, as discussed in the last section (p. 36). On cooling, the reverse occurs and the appropriate amounts of latent heat are liberated when condensation and freezing set in.

Every gramme of ice melted takes up 334 J (80 cal). Another 167 J (40 cal) per gramme will bring the water to the boil; but, a further 2257 J (540 cal) per gramme are needed to achieve vaporisation. With *gradual* warming, therefore, each gramme of water vapour produced holds about 2758 J (660 cal) of latent heat energy. Take care that the gramme-calorie and its equivalent are being used here (see Ch. 1.B). Extra energy is used when stronger temperature gradients speed up the process to result in **sublimation** (or **ablation**) and evaporation which bypass boiling. Every gramme of the resulting vapour contains 2842 J (680 cal) of latent heat.

Since the entire atmosphere probably holds close to 10^{19} g of water vapour, this represents an enormous store of energy capable of being transferred by the winds. In fact, latent-heat transfers are the most significant consequence of the cycles driven by convection (see Fig. 3.2).

In Chapter 5.A, we shall see that massive evaporative cooling by the trade winds, or tropical easterlies, removes vast quantities of heat energy from the oceans. This is released upon condensation high within the Hadley cell circulation (see p. 84). When coupled to the mechanical energy also 'stolen' by these winds in going against the Earth's rotation, the total energy released aloft powers the general circulation (see Ch. 1.G).

At a local scale, because water needs so much energy and is slow to heat up and cool down, it is said to have a high *specific heat* capacity. In fact, it is the highest by far of any known substance. A common consequence of this property is that wet materials such as waterlogged soils respond slowly to rises and falls of temperature by comparison with dry materials which are more sensitive to temperature changes. By giving up moisture, stored heat energy is removed, but the potential to absorb more and rewarm increases correspondingly. This is an important process in soil warming and plant growth, particularly in the spring. Once again we see that moisture changes regulate heat energy too; the wetting and drying of our clothes are good examples of what happens. Put simply, water is a 'carrier' of energy.

Humidity of the air
It has been seen already that the amount of water vapour in a given parcel of air defines its humidity. This can be expressed in one of three ways; its partial pressure, mass per unit volume and percentage present against what could be present at a maximum. The first two are methods of giving the **absolute humidity** because they state actual quantities of vapour. On the other hand, the latter is a proportion and so it is known as the **relative humidity** (see below).

Humidity is a function of temperature because warm air has the ability to hold more water vapour than cold air has. When a parcel contains the maximum amount of water vapour it is said to have reached **saturation point.** This occurs when the molecules no longer have the freedom to move about as required by the latent heat energy held. Any additional vapour would restrict this movement unless the parcel is externally warmed to create more room through expansion. If cooling occurs, saturation will eventually be reached as molecular mobility is restricted by contraction. Continued cooling must result in appropriate quantities of water vapour being 'squeezed out'. This entails a release of latent heat energy and so condensation sets in. With any given air parcel, therefore, saturation can be achieved in one of two ways; either by *adding* more water vapour while holding the temperature steady or by *cooling* the air while the amount of vapour remains constant. If either involve energy entering and leaving the parcel, such processes may be termed **diabatic.**

The temperature to which air must be cooled to become saturated by the water vapour it *already* holds is called its **dew point.** Diabatic principles are put to use in its measurement. If an ordinary thermometer bulb is covered with damp muslin and then given time to settle into equilibrium with the shell of saturated air it creates, a lower reading results because of evaporative cooling mentioned earlier in this section. The difference of *depression* of the *wet bulb* reading from *dry bulb* one, therefore, is a measure of the relative humidity. When they both read the same, the air is already saturated and relative humidity is 100%; therefore, the more the difference the lower is the relative humidity.

It is important to realise that air is either satu-

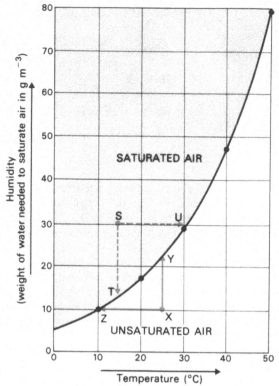

Figure 3.7 Temperature and humidity relationships

rated at 100% relative humidity and above or it is unsaturated; there are no 'in betweens'. Figure 3.7 shows a curve between saturated and unsaturated air dependent upon the maximum weight of water vapour that can be held at certain temperatures; e.g. at 10°C only 10 (g m⁻³) of air can be held without condensation taking place.

If we imagine an air parcel X with a temperature of 25°C and 10 g m⁻³ of water vapour, then its saturation can be attained either by more than doubling its vapour content along line X–Y or by cooling it to 10°C along X–Z. The second example shown is a parcel S with a temperature of 15°C and 30 g m⁻³ of water vapour. In this case considerable condensation is occurring so that over half the moisture present must be liquefied along S–T. Alternatively, warming along S–U to 31°C would re-evaporate all the condensed vapour. Other actual examples might be tested out on Figure 3.7. The formation and dissolving of clouds, fogs, and even steam in the kitchen and bathroom, are indications of these processes at work.

Diabatic effects
Figure 3.8 illustrates the typical effects of **diabatic**

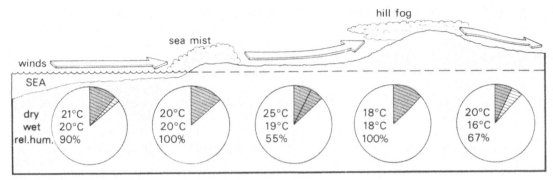

Figure 3.8 Diabatic processes along a coastline

processes along a stretch of the Cornish coastline open to strong onshore breezes on a warm summer morning. Wet and dry bulb temperatures are given for five sites and relative humidities are recorded. Only a slight overall decline in the maximum possible water vapour contents is evident going inland, as shown by the whole shaded sectors. On the other hand, the red portions reveal that localised chilling and warming have a considerable influence in producing sea mists and hill fogs. A slight drop in dry bulb temperature is sufficient to generate the mist which clings to the cliffs. This clears inland over the already warmed plateau. Further cooling on the chilled shoulder of the low hills causes a thin fog to blanket the summit, but this dissolves on the leeward slopes which are being warmed by the early morning sun. Both mist and fog dispersed by the middle of the day in question. Similar situations can be observed along most coastlines and in mountains (see Ch. 4.B and 4.C).

Other diabatic effects can be seen in association with large lakes. For instance, cold air blowing across a relatively warm lake may produce downwind showers, and warm air drifting over a cold lake invariably causes fogs. In the former, sufficient water vapour is added to induce saturation whereas, with the latter, both water vapour and chilling to dew point work together in producing the fog. Because they are the result of horizontal air flows, such occurrences may be called **advection fogs.**

It can be shown experimentally that condensation will actually set in with relative humidity as low as 78%. This is directly dependent upon the concentrations of hygroscopic nuclei present in the appropriate size range (see p. 53). These particles or aerosols attract the water and trigger the early onset of condensation. Many fogs develop prematurely and stay longer in smoky industrial regions because the water droplets adhere strongly to pollutants and create **smogs.** Indeed, water vapour itself is a product of many industrial processes, particularly the cooling towers of power plants.

If the air was pure and free of nuclei, however, supersaturation well above 100% and supercooling substantially below freezing would occur *without* condensation. Moreover, when droplets eventually begin to condense out they would be too small and most would re-evaporate long before reaching the ground. Theories to account for the observed characteristics of condensation and precipitation will be reviewed in Chapter 3.E later.

Clouds and fogs are helpful as the first visual indicators of atmospheric processes at work. However, we know now that, but for particles of the necessary size and type to act as nuclei for water molecules, there would be few clouds and little rain anywhere.

D Heat in the air

Without heat energy there could be no life. Even the most primitive peoples have concluded that the Sun is the ultimate 'giver' of this life. In Britain, for example, structures like Stonehenge show that the Sun's movements held deep significance to ways of life over 5000 years ago. In this section we shall see that little has changed fundamentally. Indeed, there is a general reawakening to the benefits of solar heating and house designs which take advantage of local climates.

Solar energy
Heat energy is the *condition* of molecular mobility in a substance. It is measured by thermometers containing materials like mercury which are quick to react to this activity. Flows of *sensible* heat

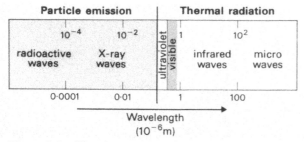

Figure 3.9 The electromagnetic spectrum

energy are achieved in the air from places with higher temperatures to those with lower ones. All molecular activity seeks to 'settle down' in this way unless stimulated to renewed action *externally*. It has been determined that, without such stimulation, all molecular activity would cease at a temperature of minus 273°C. This value is called **absolute zero** and gives rise to an alternative temperature scale named after the eminent 19th century physicist Lord Kelvin, e.g. 0°C is 273 K, 10°C is 283 K, 20°C is 293 K and so on.

Radiation energy was first explained by Maxwell in 1888. His so-called electromagnetic theory (see p. 2) envisages the energy travelling along waves with different lengths between their crests. These tiny distances are measured in micrometres, such units are simply called microns, each being 10^{-6} m. Figure 3.9 shows the main range of the wavelengths and it is worth comparing with Figures 1.7 and 1.8. It forms a continuum or *spectrum* of waves broadly banded into groups. Notice that visible light which represents Man's 'window' on

the world falls in a narrow band between about 0·4 and 0·8 microns. Above this band lie longer *infrared* waves and immediately below it are shorter *ultraviolet* waves. Together, ultraviolet, visible light and infrared waves constitute *thermal* radiation. Even shorter waves (to the left on Fig. 3.9) are the highly dangerous ones arising from particle emissions as in radioactivity. Longer waves (to the right of Fig. 3.9) are the harmless forms used to transmit radio and television signals.

The Sun may be regarded as a giant nuclear fusion pile or furnace sending out streams of radiation from its surface at a temperature estimated to be about 6000 K. Figure 3.10 illustrates that Earth receives a minute fraction via the solar 'beam' because it is only a small and distant 'speck' by comparison. The magnified drawing of Earth to the right of Figure 3.10 shows how its axis of rotation is tilted at 66½ degrees of arc to the plane of its orbit so that the beam bathes the daylight hemisphere. Note that the mean surface temperature of the Earth is 287 K or about 14°C.

There are two key radiation laws which highlight the controlling influence of temperature, i.e. the absolute temperature of the radiating surface. Both are based upon emission from perfect radiating surfaces called **black bodies.** The first or **fundamental law** states that intensities emitted from a black body surface are proportional to the fourth power of its absolute temperature. The second or **displacement law** holds that the maximum wavelength of the most intense radiation is inversely proportional to the absolute tempera-

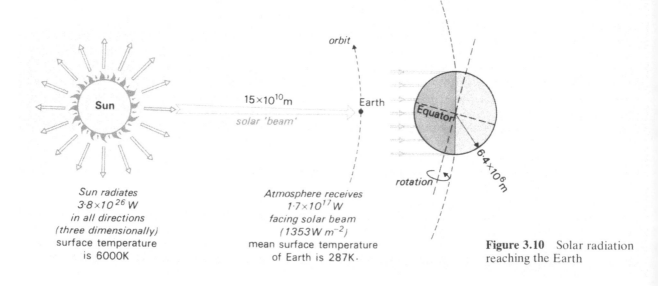

Figure 3.10 Solar radiation reaching the Earth

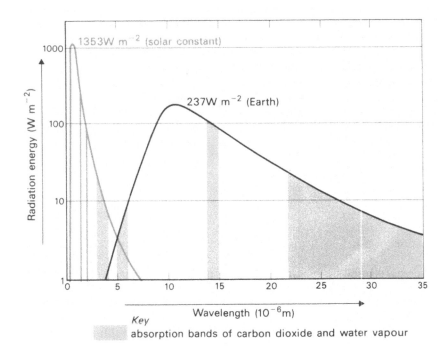

Key
absorption bands of carbon dioxide and water vapour

Figure 3.11 Incoming and outgoing radiation to the atmosphere

ture. When put together, these laws require that solar radiation will be most intense at the *shorter* wavelengths whereas Earth's radiation, assuming it to be a black body too, must be much less intense with a *longer* wavelength.

Figure 3.11 shows curves for both solar and Earth's radiation on one graph for comparison. Notice that the scale of wavelengths is divided uniformly (linear), but the scale for radiation energy increases by powers of ten (logarithmic). Such a presentation tends to exaggerate the lower radiation levels at first glance whereas, in fact, this is a false impression. Solar radiation is actually much greater and at 1353 W m^{-2} is a maximum at nearly 0·5 microns. Earth's radiation from the atmosphere to space, on the other hand, is only

237 W m^{-2} and it peaks at 10 microns. It is, of course, no accident that plants and the eyes of animals have evolved to resolve visible light around the peak for solar radiation between 0·4 and 0·8 microns. About 45% of solar radiation falls within this range, another 45% occurs in the infrared band, and the remaining 10% is contributed by ultraviolet waves. Outgoing radiation from the Earth, on the other hand, is wholly in the infrared range.

Because the Earth's annual orbit around the Sun is not circular, there is a regular annual variation in the solar constant reaching the atmosphere. We are closer to the Sun on 1 January, at $14·7 \times 10^{10}$ m, and furthest away on 1 July, at $15·2 \times 10^{10}$ m; the former is known as the **perihelion** and the latter as the **aphelion.** Figure 3.12 shows the resultant fluctuations in radiation. During the northern hemisphere's summer, therefore, solar energy is slightly reduced whereas in winter it is rather higher than the mean. This has a perceptible effect on seasonal energy budgets but it is difficult to assess because of the great contrasts in the geography of the northern and southern hemispheres.

Energy balances in the air
The apparent imbalance between incoming and outgoing radiation shown on Figure 3.11 requires explanation. At first glance, it looks as if the atmosphere is taking in more energy than it loses.

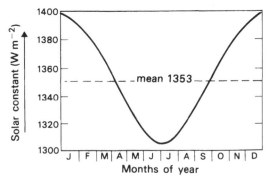

Figure 3.12 Annual variations of the solar constant

This cannot be so, of course, otherwise overheating would have destroyed everything long ago. The puzzle is resolved, first by the *geometry* of the rotating Earth and secondly by the *selectivity* of the atmosphere and Earth in accepting or rejecting radiation. Both are outlined below.

A cross section sliced through the solar beam reaching the atmosphere is consistently πR^2 (the area of a circle) where R is the radius of the Earth. On the other hand, this is spread out over all the Earth's surface area of $4\pi R^2$ (the area of a sphere) during a complete revolution in a day. Thus, the *effective* solar radiation must be *quartered*, because in any 24-hour period it is spread over a surface area 4 times larger than the receiving area. Consequently, effective incoming radiation must be reduced to 338 W m^{-2}.

On passing through the atmosphere, about 30% of incoming short-wave radiation is *rejected* because it is immediately reflected back into space from the upper surfaces of clouds, some aerosols and the ground itself. Geographically, reflectivity varies greatly of course and it may be measured by a value called the **albedo** which gives the radiation reflected as a percentage of the total received by a surface (see Ch. 4.D). Here, the 30% figure must be regarded as a rather general global value. Some authorities, in fact, estimate it to be up to 10% higher. If we assume that about 70% is *accepted*, however, then the amount of incoming radiation 'put to use' is reduced further to 237 W m^{-2}. Therefore, it is in balance with that re-radiated by the Earth (see Fig. 3.11).

Rather as a liquidity problem is created by a business which spends money as fast as it is earned, so an immediate in and out transfer in radiation would leave nothing to drive the atmospheric machine. Clearly, *banks* of energy must be held in the atmosphere. Reference back to Figure 1.6 will show that there are two such banks in the air itself, one like a savings account which *absorbs* 47% of the entire income and another like a current account which *stores* about 23% for more immediate transactions. Put simply, nearly a quarter of the funds are held for day-to-day business. Hopefully, the analogy has been taken far enough. What actually happens in the atmosphere is now considered briefly. Figure 3.13 indicates the main exchanges and balances involved, including that which is reflected and unused; red arrows show gains of radiant energy and black ones the losses.

The banking system in the atmosphere is usually called the **greenhouse effect** because of its similarity with the way that a glass covered building 'holds' heat energy.

On the income side, after the 30% has been reflected, the 10% absorbed takes out the ultraviolet waves (see Fig. 3.9). This energy goes to convert ordinary oxygen (O_2) into ozone (O_3) at heights of 20–35 km in the stratosphere. Thus, a continuous layer of ozone is sustained here to screen out otherwise dangerous ultraviolet and other short wave radiation. If this short-wave energy reached Earth, it would seriously harm all living things. The protection of the ozone screen from possible disruption by the exhaust gases of rockets, high-flying supersonic aircraft, and even the fine molecules which float aloft from manmade sprays, has greatly concerned some environmental scientists. These belong to a family of hydrocarbon 'fuels' whose manufactured name is **freons.** They contain bromine, chlorine, fluorine and iodine, which are highly reactive (see p. 36) and could 'corrode' ozone aloft.

Although amounts clearly vary, further absorption by aerosols, carbon dioxide and water vapour is relatively small. So, the atmosphere is largely *transparent* to radiation from 0·4 to about 3 microns. Since it has been seen that the ultraviolet contribution of solar radiation is about 10%, something like 60% penetration of the atmosphere is achieved by *all* energy entering the system. It is substantially true to say that the Sun's energy heats the ground *first*. Afterwards, the air is warmed from the surface *upwards*.

Radiation from the warmed lands and seas is entirely infrared above 4 microns (see Fig. 3.11). This is absorbed selectively by carbon dioxide and water vapour in the broad red bands shown. The rest is 'lost' to space directly. Because of the absorbed energy, the air itself can begin to radiate. This is called **counter radiation** and is even longer in wavelength and emitted in *all* directions (shown as wavy arrows on Fig. 3.13). Some goes on into space and is unused, but a good proportion returns to the Earth as non-directional or **diffuse sky radiation** (as distinct from the 'original' direct beam radiation from the Sun).

Returning to our banking analogy for the moment, it can be fairly argued that diffuse sky radiation is the interest on the energy saved in the air. Unlike direct solar radiation which only operates during daylight to surfaces exposed to its beam, diffuse sky radiation continues throughout the night too and gets to most otherwise shaded surfaces which rarely or never 'see' the Sun.

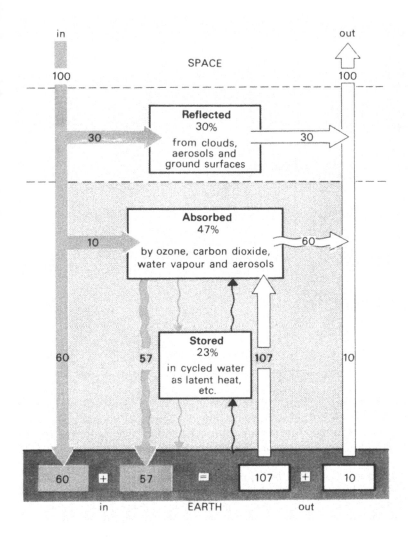

in

100

SPACE

out

100

Reflected
30%

from clouds,
aerosols and
ground surfaces

30

30

Absorbed
47%

by ozone, carbon dioxide,
water vapour and aerosols

10

60

Stored
23%

in cycled water
as latent heat,
etc.

60

57

107

10

| 60 | + | 57 | = | 107 | + | 10 |

in   EARTH   out

Figure 3.13 Radiation exchanges and balances

At this stage, we must remind ourselves of the separate energy losses and gains from the surface owing to latent heat being removed in water vapour and returned as counter radiation and precipitation. Most of the latent heat energy released on condensation aloft contributes to the absorption bank to become part of the counter radiation exchanges; the precipitation contains the mass energy largely 'drained' of its heat. So, it is now apparent that the hydrological cycle has an important part to play in building up or running down the crucial absorption bank.

Everyday weather conditions will have longer-term effects on climatic cycles of varying time scales because of the CO_2 and water vapour stores in the air (see Ch. 5.C). Just one example will suffice here. Increased warming owing to additional CO_2 and the removal of energy from both land and sea by evaporation and latent heat boosts the moisture and temperature levels in the absorption bank. This gives rise to more clouds and increases counter radiation and precipitation initially. In the longer term, however, greater reflection of solar radiation from cloud surfaces and more screening aloft reduce direct radiation to the ground. Eventually, surface cooling will cut back evaporation and latent heat losses, and the cycle is completed.

The seasons
Chain reactions similar to the one just outlined explain broad seasonal weather patterns. All are a consequence of variable surface warming and cooling globally at different times of year.

Figure 3.14 shows the fundamental reasons for these variations as the spinning and tilted Earth

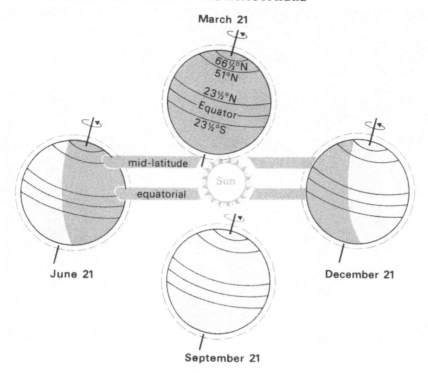

Figure 3.14 Seasonal heating

orbits the Sun every year. Four positions are given; the June and December **solstices,** when the durations of daylight and darkness vary most with latitude, and the March and September **equinoxes,** when there are 12 hours of day and night everywhere. It can be seen that these differences stem from the attitude of the axis of rotation to the plane of the orbit. Notice how night is always the shaded hemisphere. The beams of solar radiation bathing the sunlit hemisphere are parallel because of the vast distance between the Sun and Earth. Two examples are selected on Figure 3.14 to equatorial and middle latitudes respectively.

Three effects of the spinning and tilted Earth on its annual orbit of the Sun can be stressed. First, although not most importantly, any direct beam of solar radiation must pass through a greater thickness of atmosphere at higher latitudes. Secondly, any surface or horizontal plane parallel to the ground receives more intense heating per unit area in lower latitudes because the midday sun is more directly overhead. The angle made between the beam and surface is known as the **angle of incidence.** In the same way that morning and evening sunlight shines from just above the horizon and is spread broadly over the ground, so even the noon sun covers a relatively larger surface area at higher latitudes. Thirdly, owing to the tilted axis of rota-

tion, places in the northern hemisphere receive more daylight than darkness between the March to September equinoxes and vice versa for the rest of the year. The reverse pattern occurs in the southern hemisphere, of course, so that, north and south of the Equator, the seasons are always opposite.

All three effects are a function of *latitude* and the *time* of day and year. These relationships are considered further in the next chapter. Here, we only need to note that heating increases equatorward at all times and is **perennial.** Conversely, daily or **diurnal** balances gradually accumulate when the duration of daylight exceeds that of darkness going polewards. When nights are longer than days, the balances run down. The resultant **seasonal** contrasts between warmer summers and colder winters are greater at higher latitudes. Because such heat 'economies' take time to respond, the hottest period of the day is usually up to 3 hours after the most intense solar heating at noon, and the warmest days are normally a month after the summer solstice. Similar time lags are evident with both diurnal and seasonal cooling. Thus, July is usually the warmest month in the northern hemisphere and January the coldest one.

The upper graph on Figure 3.15 brings together the radiation exchanges with respect to latitudinal zones. The scale is based upon net annual totals

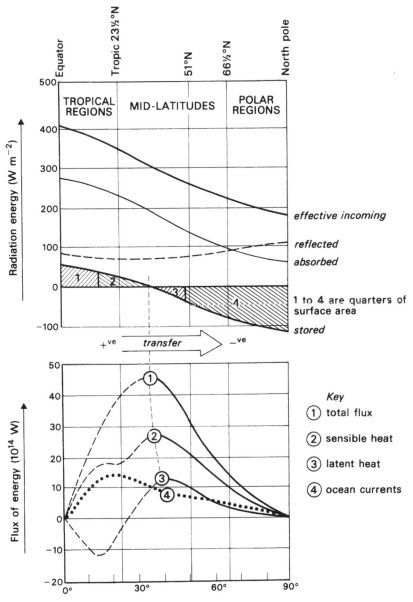

Figure 3.15 Radiation balances and energy transfers in the northern hemisphere

averaged out uniformly to indicate *continuous* flows of radiant energy. Notice the general decline of the effective and absorbed values poleward as the amounts of radiation reflected increase. The latter is at a minimum in mid-latitudes because of reduced cloud cover in the anticyclonic regions there (see p. 86). In the Arctic it is at a maximum owing to the high albedo of the ice cap. Snow and ice surfaces are highly reflective so that solar radia-

tion is returned to space largely unaltered. This severely reduces the amounts of effective radiant energy to the air above such regions.

Overall, a surplus store of energy is evident within the tropics with a corresponding deficit in higher latitudes where losses exceed gains annually. Latitude 35°N is the approximate point of balance. It is important that the areas shown for the surplus and deficit regions are actually equivalent in terms of their surface areas (see the quarter divisions on Fig. 3.15 which show that zonal areas are bigger in lower latitudes).

Surpluses from low latitudes must be transferred northwards and southwards (i.e. meridionally) to make up for deficits at high latitudes in both hemispheres. This process creates a continuous **flux** of energy poleward, as is shown on the lower graph in Figure 3.15, which is why the Tropics may be regarded as the boiler room of the atmosphere (see Chs 1 and 2). Different curves show the main components of the total flux. These are drawn as broken lines in the surplus regions and continuous ones in the deficit zones. It is worth noting that ocean currents account for up to a quarter of the energy moved out of the tropical regions. Broadly, the circulation of the oceans matches that in the air. Also, it can be seen that the removal of latent heat is only extra-tropical as the Hadley cell circulation moves water vapour at low levels *into* the Tropics rather than away from them. All maximum fluxes in the atmospheric circulation occur in mid-latitudes and this reflects the importance of air mass exchanges there, as was explained at the end of the last chapter. These are seen to be close to the latitude of the British Isles. Finally, it should be mentioned that graphs for the southern hemisphere would be similar in most respects.

Once again a distinction can be made between the Tropics, which need imports of moisture or mass energy, and the extra-tropical zones, which rely upon heat energy from elsewhere. Returning to where this section began, it is little wonder that people in the Tropics have ways of life which stress the availability of water and rain while heat and the Sun have a more obvious impact upon lives in temperate latitudes.

E   Stability of the air

By continually striving to distribute energy uniformly, the horizontal fluxes seek to establish a *stable* atmosphere globally which would be barotropic (see p. 19). Large **air masses** with widespread uniformity of humidity and temperature in layers are the closest to this goal. Within them, however, much smaller **air parcels** must move up and down, particularly when the air mass passes over surfaces with irregular relief and locally variable heating. In this concluding section, the **stability** of the smaller parcels with respect to their stable surrounds is considered. Here we see the local interplay between moisture and heat energy causing our weather.

Adiabatic processes

The properties of large stable air masses are regarded as being conservative in that they do not change readily having been slowly acquired in their source regions diabatically (see pp. 40–41). This may take several days. Figure 3.16 shows the main **source regions** for tropical and polar air masses. When compared with Figure 2.25, it generally indicates that the mid-latitude anticyclones are the 'breeding grounds' for the **tropical air** masses (T) while the scope for **polar air** (P) is much more confined. The former often persist for weeks on end substantially unchanged. Air masses which develop over land are termed **continental** (c) and those above the sea as **maritime** (m). Notice that the leading boundaries of the contrasting tropical and polar air give rise to **warm fronts** and **cold fronts** respectively. They will be considered in more detail in the next chapter in the context of depressions and anticyclones crossing northern Europe. Here, attention is drawn to the marginal position of the British Isles which means that they are influenced by any of the four types of air mass shown.

The general situation illustrated on Figure 3.16 typifies winter when the star-like polar air in the middle is largest and more dominant. Remember that Rossby waves are really responsible for this shape and that the whole system slowly revolves with the Earth's rotation (see p. 30). In the heart of the polar vortex above the ice cap lies **arctic air** (A) and around the circumference of the map lies a belt of **equatorial air** (E) (neither are shown on Fig. 3.16).

Once a small parcel of air is moved away from the ground surface, it is cut off from any supplies of moisture and heat energy and 'on its own'. Moreover, it will almost certainly have moved into another layer of the surrounding or **environmental** air mass. The conservative behaviour of the parcel means that it becomes a discrete bubble or unseen balloon of air which will not mix with its environment. Since its properties can only change *internally*, therefore, the processes governing its subsequent behaviour are called **adiabatic,** which simply means 'impassable'. This contrasts with the 'passable' diabatic transfers already explained (see p. 40) which many prefer to call non-adiabatic anyway. Furthermore, adiabatic processes are vertical rather than horizontal.

As an air bubble rises and moves into regions of lower pressure it expands. Even this requires some work, however, which reduces internal molecular

Photograph 1 Clouds above localised thermals. Typical fair weather cumulus clouds above warmed surfaces during fine sunny afternoons. Small bubbles of humid air are lifted beyond condensation level. The photograph was taken in July over Northern Ireland from an aircraft flying at about 1000 m.

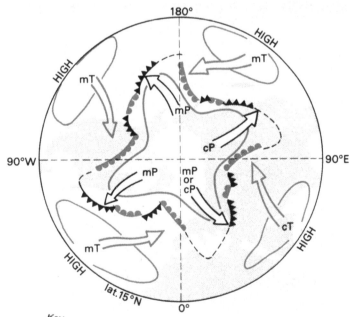

Key

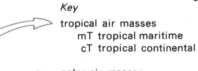

tropical air masses
   mT tropical maritime
   cT tropical continental

polar air masses
   mP polar maritime
   cP polar continental

warm front
(leading boundary of
  tropical air)

cold front
(leading boundary of
  polar air)

Figure 3.16 Air masses and fronts in the northern hemisphere

activity and leads to a fall in temperature. Thus, cooling results from lifting and adiabatic expansion. Since an unsaturated bubble will also hold certain amounts of water vapour, any decrease in temperature must raise its relative humidity (see Fig. 3.7 and the X–Z line shown). If enough lifting and cooling take place, the temperature approaches dew point, the bubble becomes saturated, and condensation sets in. This sequence will be accelerated by the presence of hygroscopic nuclei. Photograph 1 shows bubbles that have condensed into clouds.

While the bubble remains unsaturated, it can be shown that its rate of cooling owing to expansion is a consistent 10°C for every 1000 m ascended. This is called the **dry adiabatic lapse rate (DALR).** Once saturation is achieved, however, the gradual release of latent heat with condensation reduces the rate of cooling to the more variable **saturated adiabatic lapse rate (SALR).** A rule of thumb is that for temperatures above 30°C the SALR is about 5°C per 1000 m rise, down to freezing point it is around 6°C per 1000 m, and below 0°C the rate climbs to above 7°C per 1000 m.

On the basis of both adiabatic lapse rates, it is possible to calculate the approximate height of a cloud base from wet and dry bulb temperatures taken at the ground surface. For example, with a dry bulb temperature at 20°C and a wet bulb reading of 10°C, the difference would be attained in a 1000 m ascent dry-adiabatically. Cloud base might be slightly lower, of course, given many condensation nuclei (see p. 41).

Air mass lapse rates

Figure 3.17 illustrates two cases of bubbles cooling adiabatically. The one over the hills might be lifted *mechanically* in the first instance and the one above the hot valley could be a convective bubble breaking loose *thermally*. Isotherms for the environmental air mass are given. Because of contrasting initial temperatures and humidities at the surface, different cloud bases are evident. It is also apparent that both parcels remain slightly warmer than their environmental surrounds while cooling adiabatically. This ensures that they carry on rising well above the heights to which the initial lifting mechanisms are effective.

In the examples given, the bubble above the hills comes into thermal balance or equilibrium with its surrounds at about 3°C, some 2500 m above sea level. This stops the ascent as the bubble is no longer buoyant thermally and has been 'neutralised'. The result is a layered or **stratus** cloud with a flat upper surface. Meanwhile, the bubble above the valley remains warmer than its surrounds and even more buoyant after condensation level. Consequently, it continues rising on its own accord, cooling along a saturated adiabatic lapse rate until freezing begins at about 3500 m. Such unrestricted billowing gives rise to a towering **cumulus** cloud.

Figure 3.18 expresses the information shown on the previous illustration as two graphs, one for the stratus cloud over the hills and another for the cumulus cloud above the valley. Each one shows the adiabatic lapse rates (DALR and SALR) and **environmental lapse rates (ELR)** as tempera-

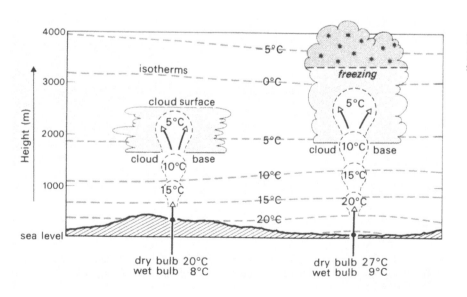

Figure 3.17 Adiabatic processes above different surfaces

Photograph 2   Billowing clouds. The 'boiling' upper surface of a deep layer of cumulus clouds developed in unstable air near Iceland. (Seen from an altitude of about 5000 m)

Photograph 3   Layered clouds. The flat upper surface of a widespread bank of stratus clouds developed in stable air over the middle of the Atlantic Ocean. (Seen from an altitude of about 10 000 m)

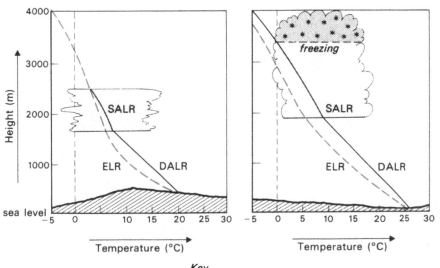

Figure 3.18   Adiabatic and environmental lapse rates

Key
DALR   dry adiabatic lapse rate
SALR   saturated adiabatic lapse rate
ELR   environmental lapse rate

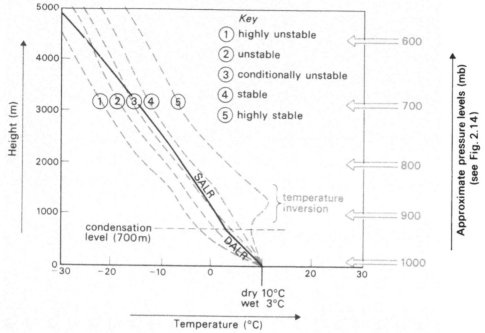

Figure 3.19 Air mass stability

ture–height curves. The former will always have the same slopes, of course, but environmental lapse rates will vary with different types of air mass (see p. 48). Note how the adiabatic curves meet the environmental one at the top of the stratus cloud while those for the cumulus cloud remain to the *right* of the ELR on the *warmer* side. Photographs 2 and 3 show examples of these cloud types.

The relationships between adiabatic and environmental lapse rates on such tempera-ture–height graphs form the basis for determining air mass stability. In addition, they explain cloud types and heights and give an insight into the processes causing precipitation. Not surprisingly, therefore, graphs similar to those on Figure 3.18 (called **tephigrams**) are important tools in weather forecasting. The information upon which environmental lapse rates are prepared onto tephigrams is obtained from balloon and rocket ascents into the upper air (see p. 11).

Figure 3.19 shows five separate environmental lapse rates on one graph. All have surface dry-bulb temperatures of 10°C and wet-bulb temperatures of 3°C and so start at the same place on the ground. Given sufficient lifting, therefore, condensation would occur at 700 m above sea level. The adiaba-tic curves (DALR and SALR) are also the same for any one of the five cases shown. Notice that

these lie to the *right* of both the unstable air mass examples (i.e. always warmer) but are to the *left* of the two stable cases (i.e. always colder). Number 3 in the middle hovers closely either side of the adiabatic curves, typifying the all too common situation of **conditional instability** which bedevils much forecasting in the British Isles.

In unstable air the rates of adiabatic cooling are always less than the environmental lapse rate. In the first case (1 on Fig. 3.19), any mechanical or thermal lifting of bubbles from the surface will cause unrestricted ascents and deep cumulus clouds well into freezing levels. In the second case (2 on Fig. 3.19), the growth of such clouds would stop just above 4000 m because thermal equilib-rium exists there. With the third situation (3 on Fig. 3.19), showing conditionally unstable air, it would be necessary to lift air bubbles, probably mechanically over mountains, well over 1000 m before they could rise to about 3500 m on their own. Similarly, unless, lifted by independent means above condensation level, both stable air masses (4 and 5 on Fig. 3.19) would have cloudless skies. Any bubbles forced to rise from the surface would sink back again as soon as the lifting ceased. This is notably so in the case (5 on Fig. 3.19) showing an inversion of the normal decrease in temperature with height (see p. 44).

Precipitation

Considerable potential energy is given to clouds resulting from air mass instability in terms of both the weight of liquid *water* present and the amount of latent *heat* released aloft with condensation. The liquid water will come to Earth as precipitation when the droplets become heavy enough to attain sufficient falling speed and overcome their supporting updraughts. Every falling droplet is rather like an individual trying to move against a crowd going in the opposite direction. Clearly, condensation alone is inadequate, as is evident in the swirling droplets in a fog or mist.

Following much research with the help of aircraft, two theories which explain different forms of precipitation have been advanced. The first is attributable to the pioneer work of the Norwegian meteorologist T. Bergeron in 1933 and the second to Sir George Simpson and B. J. Mason in Britain a little later.

The so-called **Bergeron process** requires freezing temperatures down to below minus 30°C in some instances. Tiny ice needles and crystals grow around minute nuclei by sublimation (see Fig. 3.6) at the expense of any supercooled water droplets present, i.e. vapour evaporated off the droplets turns to ice. These ice particles 'knit' into loose flake-like crystals in very cold and calmer conditions, but become compact pellet-like 'stones' in more turbulent air as droplets attract layers of ice by accretion (see p. 39). The two forms are known as **snowflakes** and **graupel** respectively. If neither melt while falling through the warmer air layers below, the former arrives as **snow** and the latter as **hail.** The turbulence required for hail makes it occur more frequently in summer in association with deep convective stormclouds. When melting does occur, both give rise to large **raindrops** and, maybe, heavy downpours in the case of those supporting large graupel aloft. **Sleet** is an intermediate form of snow and rain in near-freezing surface air. Bergeron processes, therefore, are favoured by localised convectional conditions and the activity along frontal troughs in temperate depression systems. Both are considered in the next chapter. Suffice it to mention here that the resultant forms of precipitation at the ground largely depend upon the stability of the air and its lapse rate. Rain or hail occur more with unstable lapse rates above warmed surfaces in summer whereas sleet or snow can occur in more stable air over cold surfaces in winter.

The **Simpson and Mason's theory** accounts for the exceptions to the phenomena described above, with particular reference to tropical rainfall which does not begin with freezing aloft. 'Giant' hygroscopic nuclei are thought to produce big unstable liquid water droplets by collision processes. Huge salt particles above evaporating and rough seas are ideal in attracting water vapour which then condenses into droplets of varying sizes. If sufficient cloud depth is attained, say over 1 km, there is opportunity for any fast-moving small droplets to hit and be captured by the slow-moving large ones. Droplets may begin as tiny globules only 0·1 mm in diameter, then grow to maximum sized raindrops of 5·0 mm in diameter. Thereafter they would burst themselves. Without the initial giant nuclei and larger droplets, however, uniformly small sized droplets do not appear to collide, grow and become heavy enough to fall. This is evident in fine mists and drizzles with droplets less than 2 mm across.

Other forms of precipitation such as **dew** deserve mention since they are important to plant growth, notably in dry climates. Just as water vapour in a room condenses against such chilled surfaces as windows, so dews form on thin blades of grass and leaves which quickly lose their heat at night (see Table 3.2).

Like the end products of a long and involved assembly line, clouds and rain make a fitting and familiar conclusion to this chapter. Furthermore, their great variety tell us much about the manufacturing processes involved. The reader is invited, when it rains, to reverse the forecasting procedure and list, with the benefit of hindsight, the chain of events involving both moisture and heat that has led to the weather concerned.

Chapter 4

Local Processes

The weather on a nearby hillside or in the street outside is a blend of atmospheric processes brought together from distant parts of the world and locally modified. So, the study of local processes is more meaningful within the broader framework of the global circulation and synoptic situations already discussed.

In this chapter, each section in turn explores processes ever closer to home. Although selected case studies are used, the threads of moisture and heat energy will be strong in them all.

A Pressure systems crossing Britain

Summaries of the weather associated with the passage of typical low pressure systems or **depressions** and high pressure systems or **anticyclones** (see Fig. 2.5) over the British Isles will serve as key stones in the bridge between the previous chapter and this one.

Low pressure
Figure 4.1 shows a depression system passing over north-western Europe with special reference to a line of places across the British Isles into the Baltic. Each synoptic map or surface pressure chart in the sequence is accompanied by a section of the troposphere along the line. Different aspects of the weather are shown on a separate section, i.e. winds, temperatures, clouds and rainfall. The last one includes an inset graph showing the accumulation of rainfall during the day at the third station in South Wales. Although a day characteristic of either spring or autumn has been chosen, small increases or decreases in temperature would make the situation apply to any time of year. The *sequence* of weather accompanying such systems and outlined below is much the same in all seasons.

In the case shown, a depression centre passes northern Scotland towards Scandinavia during the day. Fronts (see p. 19) associated with the different air masses drawn into the low- from the higher-pressure regions in mid-latitudes extend across the British Isles and weaken southwards. They move as north–south belts of changeable weather across the country, the cold front gradually catching up the warm one until meeting and even 'overtaking' it in the North Sea. When this happens the fronts are said to have **occluded,** which simply means to shut or close up. Notice that air pressure in the centre of the low *deepens* to 980 mb as the occlusion begins. Thereafter the low *fills* as the warm sector of tropical maritime air is squeezed upwards. This is an indication that the system is decaying or breaking down at the end of its short life. Thus, occlusions almost literally 'zip up' the system.

Such depressions frequently form off the coast of Labrador (see Ch. 2.G), then are driven and steered rapidly across the North Atlantic towards Europe ahead of the upper trough and jet stream. They mature over the ocean, being fed by warm tropical maritime air and drawing in cold polar air in their wake. These are spun and pulled upwards within the core of the depression (see Fig. 2.23).

Before it arrives, Britain is influenced by a dry and stable tropical continental air mass drifting from south to north across Europe. In the afternoon, daytime heating would be sufficient to produce small 'fair weather' **cumulus** clouds above warmed surfaces; but, these would dissolve by late afternoon (see Photo. 1). During the night, however, clear skies would be evident. By 0600 hours surface pressures begin to fall with the winds slowly **veering** or going clockwise more from the south-west. They will also strengthen in exposed areas.

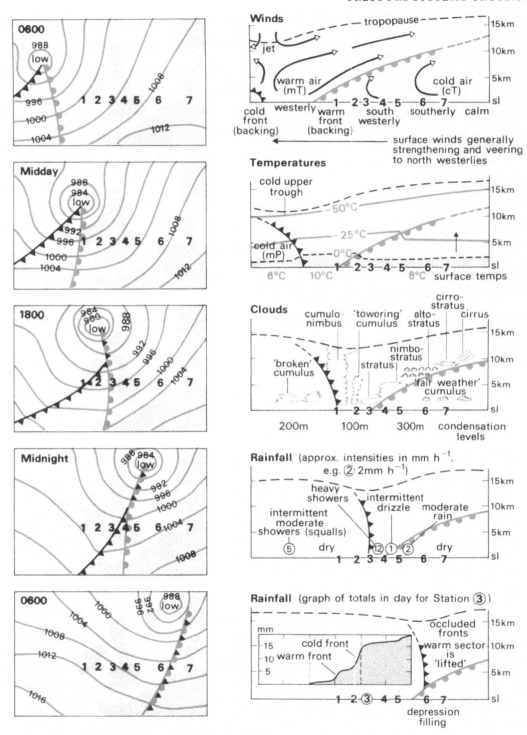

Figure 4.1   A depression crossing the British Isles

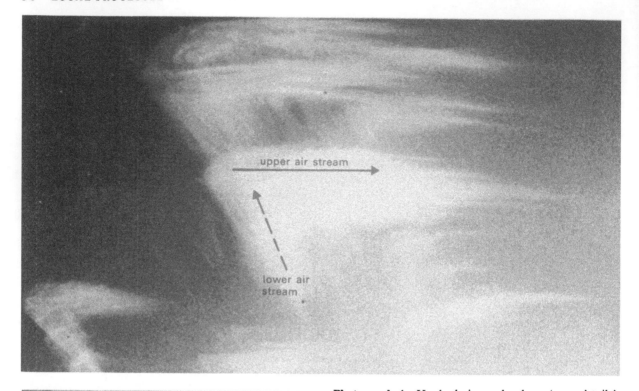

upper air stream

lower air
stream

Photograph 4 Hooked cirrus clouds or 'mares' tails'. High altitude clouds heralding the leading upper edge of a warm front and showing wind shearing aloft as air-streams blow in different directions at various levels. The clouds consists of ice crystals which 'stream out' with the winds. (Photographed from the ground)

Photograph 5 Mottled alto-cumulus clouds or 'mackerel sky'. Intermediate altitude cloud bank along a frontal surface with rolling air where winds shear at two levels (see Fig. 4.8). The lack of a distinct banded texture indicates that the directions of the lower and upper winds are variable. (Photographed from the ground)

Dawn is accompanied by high-level wisps or filaments of **cirrus** clouds, popularly called 'mares' tails'. These herald the leading edge of the warm front aloft and are often swept into curves or hooks as the upper air streams turn on their northbound run (see Photo. 4). Twilight penetrating a thick layer of the atmosphere obliquely may have sufficient short waves scattered to turn the clouds red. We shall now see that this justifies the old adage, 'Red sky at morning, shepherd's warning'.

Later in the morning, clouds thicken and become lower, often in a classic sequence. After the wisped cirrus, a hazy veil-like layer of **cirros-**

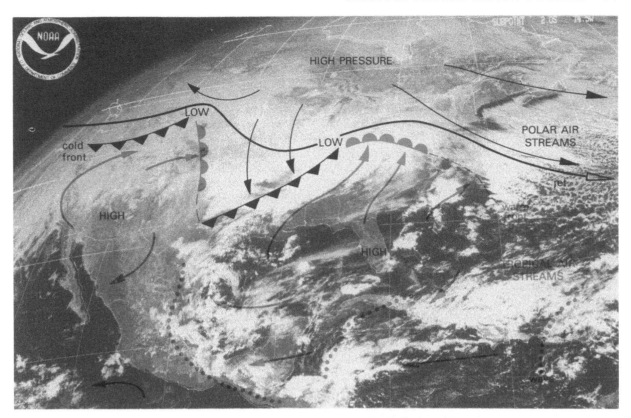

Photograph 6 Satellite picture of depression systems and high pressure regions in the Northern Hemisphere. (By courtesy of the National Oceanic and Atmospheric Administration, United States Dept of Commerce.) This photograph was taken at 1700 hr GMT on 12 February 1979 from a stationary weather satellite (GOES) some 35 800 km above the Equator. It shows the cloud cover over North America and the Caribbean from the polar to tropical regions. Such photographs are transmitted every 30 minutes.

The main features of the associated surface weather systems are also indicated on the photograph: two mid-latitude depressions or lows have formed beneath the upper jet stream blowing between the high pressure areas dominating the polar and tropical regions (compare these with the general circulation shown on Fig. 2.28).

(a) The most developed depression lies south of the Great Lakes and another is forming over Yellowstone to leeward of the Rockies (see Figs. 2.24 and 2.25). Thick clouds are swirling into both systems. Notice how the clear dry and cold polar air stream over New England becomes cloudy above the warm Gulf Stream offshore en route for Europe.

(b) Subtropical high-pressure systems are giving fine weather over Florida and the desert regions of the South West and Mexico. Further south, less organised cloud belts indicate the position of weak waves in the tropical easterlies (see Figs. 5.1 and 5.2). Notice how these form at the Caribbean islands and develop over the mountains of Central America in the induced trough between the two subtropical highs or anticyclones.

tratus covers most of the sky. This consists of small needles of ice and supercooled water droplets which filter sunlight or, maybe, even refract it to produce bright rings of light or haloes. Greater upper turbulence follows with winds at different levels blowing in varying directions. Such **shearing** creates a corrugated frontal surface characterised by patterned or banded **altostratus** cloud banks (see Photo. 5 and on to Fig. 4.8). The effect is often called a 'mackerel sky' and gives a last warning that the arrival of the warm front is imminent. Fronts are also evident on satellite pictures of clouds (see Photo. 6).

Just before the passage of the warm front, winds freshen and temporarily **back** or turn anticlockwise. Soon, a belt of low-level rainclouds called **nimbo-stratus** passes over giving moderate precipitation, especially on high ground. A noticeable rise in temperature is felt as the front moves through.

Behind the front, the tropical maritime air brings muggy weather with low uniform **stratus** clouds, hill fogs and occasional light drizzle. Localised clearances of this blanket may occur later in the afternoon as the wind veers westerly and gusts more over hills. By the evening, deep cumulus clouds build upoce ahead of the rapidly approaching cold front. Towering **cumulo-nimbus** rainclouds may be evident with their tops sliced or streamed off by a strong upper jet stream (see Photo. 7). The evening sunlight refracted through the ice particles at the edges of the billowing mass may show the famed 'silver lining'. All this indicates considerable instability aloft. Because they reach the deeply frozen roof of the troposphere and are extremely turbulent, such clouds contain all the ingredients for the Bergeron process. Heavy thundery showers result with outbreaks of hail, sleet and even snow on colder uplands.

As the cold front closes, winds temporarily back again becoming squally and driving heavy rainstorms across the countryside. With its passage there is a sharp drop in temperature accompanied by a sudden veer in the wind and a short 'clearing up' storm. The fresh or biting polar maritime air mass that follows is highly unstable because it has been warmed from below when passing over the Gulf Stream and North Atlantic Drift. Broken cumulus clouds with ragged edges scud across the sky. The familiar forecast phrase, 'to be followed by scattered or occasional showers and bright intervals' aptly sums up the weather at the end of the system. Scattered indicates a patchy distribu-

tion in *space*, and occasional refers to irregular events in *time*.

The inset rainfall graph on the last section of Figure 4.1 shows how the day's rainfall accumulates to a total of 20 mm in South Wales. Such graphs are based upon continuous rainfall records from automatic or autographic gauges and called **mass curves.** The passage of the warm front at 1800 hours and the cold one at midnight is clearly marked by steepenings of the curve. Much the same pattern would apply to other stations although totals would most likely be less further east and the times clearly later.

Finally, we can return to the occlusion which marks the end of the system's life. Occlusions occur frequently over Britain particularly since the more gently sloping warm front is held back by rough surface relief while the barely reclined cold front drives through, merely buckling to a more vertical position with dragging at the ground. Just as a toe driven hard beneath a light ball will 'chip' it upwards, so the overtaking cold front elevates the warm tropical maritime air ahead. Usually, the toe or wedge of cold air *over-rides* the warm front surface giving rise to a **warm front occlusion** (see Fig. 4.1). Sometimes, however, it pushes *beneath* the warm sector at the ground surface even more vigorously as a **cold front occlusion.** Generally, the former tends to occur in winter when the continental air ahead of the system is colder than the polar maritime air that follows (see Ch. 3.E). Rarer cold-front occlusions are occasional summer events when relatively colder and denser polar air drives hard beneath much warmer and lighter continental air ahead.

The mass lifting of humid tropical air in occlusions is accompanied by short and sharp showers. In summer, especially, these will be thundery and may even be locally *violent,* which means that falls or intensities exceed 50 mm per hour. Officially, *heavy* showers are between 10 and 50 mm per hour. If less than this, the intensity is classed as *moderate.* So the dying throes of depressions are often dramatic final 'gestures' as stores of potential energy are released as precipitation. Indeed, most depression systems undergo a troublesome birth, vigorous life and untimely death as essential ephemeral disturbances mixing tropical and polar air. Many varieties exist, therefore, which is why Figure 4.1 illustrated a composite and somewhat stereotyped version rather than a real one. Specific cases are mentioned later, but the best examples are to be found out-of-doors!

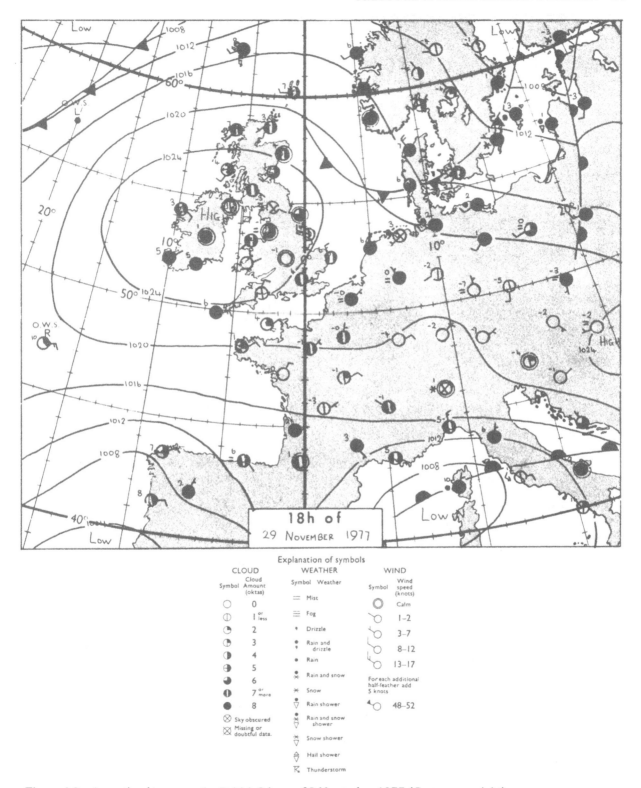

Figure 4.2 An anticyclone over the British Isles on 29 November 1977 (Crown copyright)

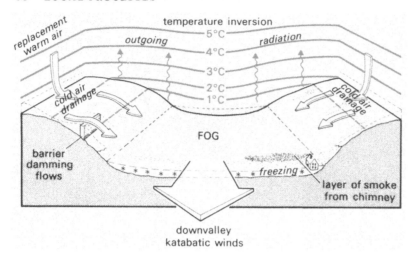

Figure 4.3 Night-time cooling and a temperature inversion in a valley

High pressure

Figure 4.2 shows an actual example of a less common anticyclone over the British Isles. This stable system persisted well into the first week of December 1977 before slowly drifting towards Scandinavia. Notice that most of the British Isles has sea-level pressures above 1024 mb with calm or only light outblowing winds. An upper ridge in a poleward loop of the Rossby wave lies to the south-west (see Ch. 2.G and Fig. 2.23). The core of subsiding air slowly warms adiabatically, reversing the processes outlined at the end of the last chapter. Thus, although low clouds exist in coastal districts, the day is clear and sunny inland especially in the southern half of the country.

When such systems occur during summer, hot fine weather is experienced. Heatwaves develop if the pressure remains high for several days. After about a week, however, the build-up of surface heating causes thunderstorms (see Ch. 4.D). Late spring and autumn are times when anticyclones frequent northern Europe with the threats of frosts and fogs respectively (see Ch. 5.B). May and even June in such temperate zones are risky times for growing plants, and the late November and early December period exemplified on Figure 4.2 sees the onset of winter. Their common processes are explained below.

The calm conditions and absence of cloud cover in an anticyclone indicate a more barotropic atmosphere (see Ch. 2.D) with well developed air layers near the ground. Direct incoming and outgoing radiation increases relatively at the expense of diffuse sky radiation (see Ch. 3.D). Therefore, when nights are longer than days (see Fig. 3.14),

outgoing radiant heat energy passes unhindered into space with little in return. Gradually the ground chills, then the skin of air against the surface and each successive layer above in turn. Even conduction (see Table 3.1) contributes a little to the heat losses in well stratified air. Soon a **temperature inversion** becomes established in the lower air layers.

Figure 4.3 shows how inversions near the ground give rise to **cold-air drainage** off slopes into low-lying sheltered valleys and basins. Sometimes

Photograph 7 Deep clouds. Cumulo-nimbus or thunderstorm clouds showing a deep condensation layer towering to the roof of the troposphere owing to strong updraughts. The flat upper surface is caused by the temperature inversion at the tropopause (see Fig. 2.13), and the anvil-like shape is attributable to a jet stream. Note the billowing clouds below which indicate vigorous convection and instability near the ground. (Photographed over southern England on a hot summer afternoon)

Photograph 8 Shallow fogs. Radiation or inversion fog showing a shallow condensation layer subsiding into a valley owing to light downdraughts or cold air drainage at night. The flat upper surface of the fog is caused by the temperature inversion in the valley (see Fig. 4.3). It is beginning to 'burn off' and lift with the morning sunshine. Note the clear skies overhead indicating stability aloft. (Photographed over northern Spain in the Cantabrian mountains on a cold autumn morning)

both temperature and slope gradients are strong and steep enough to generate cold downvalley winds known as **katabatic flows,** simply meaning 'to go down'. Cold-air drainage and katabatic winds are often cyclic as warm air needs to be drawn onto the shoulders of the valley and chilled before sinking down hill. In some circumstances the slopes remain warm while the cold air pool in the valley floor gradually deepens with radiation losses. The analogy with drainage is appropriate because the flows behave very much like water; ponding and overspilling occurs at walls and fences along contours, and such minor relief features as hillside hollows become channels for more concentrated 'rivers' of air.

The forecast warning, 'Frost is expected in sheltered low-lying districts inland', is particularly feared by market gardeners in late spring when crops are blossoming. Their steps to counter cold-air drainage repay study: warmer slopes are preferred, paraffin convector heaters and even electrical soil warming are used with large fans to stir the air and prevent its stratification. Traditional methods are to cover the crop with a *mulch* such as straw or bracken to act as a blanket. It has been found that controlled irrigation helps indirectly because the increased water vapour near the ground absorbs relatively more outgoing radiation and returns some as diffuse counter radiation throughout the night. Furthermore, the release of latent heat with heavier dews and the condensation of mists is beneficial.

Because most urban areas are situated in valleys and lowlying basins, smoke and fumes are trapped beneath the inversion to produce fog and smog blankets. This typifies the anticyclone in the late autumn, e.g. notice the freezing fogs already formed at 1800 hours on 29 November 1977 in the industrial regions of central Scotland and south Lancashire (see Fig. 4.2). Later in the night, they built up over the Midlands and London area too as regions further south chilled off. Even during summer, however, low-level inversions over cities produce palls of hazy fumes, giving rise to what is sometimes called 'anticyclonic gloom'.

Low-lying winter fogs and smogs may become so thick that they are not easily 'burnt off' or evaporated by the Sun the following day. This is helped by the low angle of the solar beam and the high albedo (see Ch. 3.D) of the fog surface (see Photo. 8). Therefore, such treacherous conditions prevail while the anticyclone lasts. Low-lying sections of motorways, including stretches in deep cuttings,

are especially prone to persistent freezing fogs. Even trapped exhaust fumes become more concentrated and thereby aggravate the hazard since most contain hygroscopic particles which 'thicken' the density of water droplets and reduce visibility (see Ch. 3.B and 3.C). Measures to restrict smoke and exhaust fumes in urban areas have staved off the worst effects of winter inversions, not least with respect to general health and serious chest ailments. To contrast them with the advection types of fog mentioned earlier (p. 41), the anticyclonic variety just described is often classed as a **radiation fog.**

The number of professions basing their operations upon regular weather briefings and using synoptic maps is impressive. We need look no further than to air lines and shipping lines to appreciate the impact of pressure systems upon our lives.

B Processes over hills and mountains

It is often said that hills and mountains make their own weather and turn bad conditions to worse. In this section, examples largely supporting these views are considered.

Local weather in uplands
By and large, the *altitude* and *amplitude* of hills and mountains determine the variations of lower-air processes. The former simply refers to the summit heights and the latter is a measure of the difference between the ridges and adjacent valley floors. They indicate how strong the relief and gradients are. Alpine terrain, for example, is clearly both higher and stronger than the most rugged landscapes in the British Isles. It follows that the inter-relationships between the processes

Photograph 9 Mountains above clouds. Clouds at 2000 m enveloping the valleys in the Picos de Europa massif, Spain. Notice how the chilled air has subsided into the valley below the inversion level (see Fig. 4.3). (Photographed on a cold summer morning)

Photograph 10 Clouds above mountains. Clouds at 3000 m capping the mountain ridges in the French Alps on the Mont Blanc massif. Notice how the warmed air above the valley is cloud free because of dry updraughts (see Fig. 4.4). (Photographed by Phillip Romford on a warm summer afternoon before visibility deteriorated on the ridge)

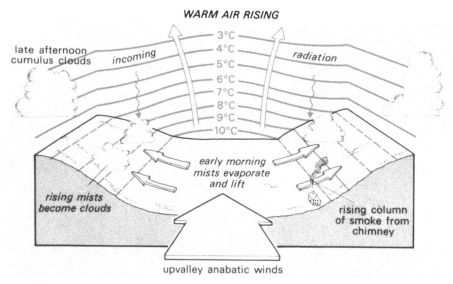

Figure 4.4   Day-time warming and thermals in a valley

in valleys and on ridges are the key ones in hill and mountain weather.

In the last section, we saw that low temperature inversions create valley floor mists and katabatic winds. This typifies alpine valleys where high snow and ice chills stable air quickly as soon as the Sun sets. Long before this happens on the mountainside, the nearby incised valley will have cooled already owing to earlier shading. With clear skies and 'thinner' air at higher altitudes, the contribution of diffuse radiation is minimal. Dawn at high elevations is often greeted with the splendid spectacle of the first warmed snow-capped peaks thrusting through a carpet of clouds still clinging to the colder valleys (see Photo. 9).

As the morning Sun warms and evaporates the upper surface of the cloud, the whole bank gradually thins out and begins to disperse. By midday the nocturnal inversion is broken down and convection currents cause strong updraughts or **thermals** which drive lingering clouds away or upwards. These generate an upvalley or **anabatic wind** during the afternoon. Figure 4.4 shows this wind and the cumulus clouds that billow and spill onto the ridges This deterioration in the weather at high altitudes as evening approaches explains the wise alpine tradition of starting to climb at no later than daybreak (see Photo. 10).

Apart from the local thermal conditions creating valley winds, passing pressure systems at a synoptic scale can induce long corridor-like troughs to become **wind lanes.** The Rhône valley between the Massif Central and French Alps has this wind-tunnel effect when the damaging Mistral blows down it. As a low-pressure system moves across the western Mediterranean towards Italy, notably in spring, cold dense air off the snow-bound uplands of central France sinks into the Rhône valley and is sucked southwards into the depression. The situation shown on Figure 4.2 is a good case in point, even though it is in November. When this happens in spring, the effects upon the *primeurs,* or early vegetable growing regions, of Mediterranean France can be devastating. It is little wonder that such farms are located to protect them from the worst damage by the Mistral.

Photograph 11   Orographic rain clouds. Unstable air off the north Pacific Ocean produces rain clouds immediately upon lifting above the Coastal Ranges of northern California. Heavy snow falls also occurred on the higher mountains above 2000 m inland.

Photograph 12   Rain shadows. Clearing skies above the intermontane plateau to leeward of the Sierra Nevada in California. The mountains are over 4000 m high and Owens Lake on the plateau is at about 1000 m. Arid conditions give rise to desert landscapes. High evaporation produces a salt lake and strong afternoon heating creates vigorous thermals which pick up dust devils off the lake surface in the distance (circled).

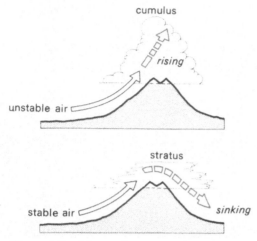

Figure 4.5   Air crossing mountains

Big mountain barriers not only deflect and diverge air streams crossing over them (see Fig. 2.1) but also lift air parcels within the mass as a whole. These parcels behave adiabatically in relation to the environmental lapse rate of the air mass (see Figs 3.18 and 3.19). Two situations can be envisaged: one with unstable air and the other with stable conditions. Both are shown on Figure 4.5.

With unstable air masses, the windward slopes literally 'kick off' the parcels so that they rise spontaneously well above the peaks. Strongly developed cumulus clouds form at or even ahead of the mountain range (see Photo. 11). Soaring currents aid precipitation called **orographic rain,** simply meaning mountain-forming. Most will fall on the exposed windward slopes.

On the other hand, stable air lifted and cooled adiabatically may only give rise to stratus clouds around the summits. These soon clear to leeward as the stable parcels sink back after climbing over the range. Little if any rain falls, except light drizzle just below cloud base or within the condensation levels themselves.

Because leeward downdraughts warm adiabatically and 'dry out', the resultant winds can be highly evaporative. Good examples occur around Geneva in Switzerland which even melt low-lying winter snows. Thus, the local term **föhn wind** has been borrowed to describe such lee winds anywhere. The cut-off tail of stratus clouds standing above these downdraughts is referred to as a **föhn wall.** It gives a clear warning to pilots flying upwind across mountains to go high above the cloud; otherwise they might be dragged down against the mountainside.

Other well-known instances of föhn winds frequent the eastern slopes of the Rocky Mountains onto the Prairies, particularly Alberta Province in Canada. Here they are named **chinook winds** after local Indian tribes. With the passage of a deep depression across the Prairies and pressure-induced downdraughts in its wake, the warm chinook can raise winter temperatures from their normal frigid −20°C to a relatively balmy 5°C in a few hours. It melts snow, creates avalanches and causes unwanted floods. In summer the chinook is equally severe because it dries out and even dessicates soils and crops. Versions of this wind further south brought about the infamous Dust Bowl havoc of the 1920's in the marginal western States.

Regional effects beyond uplands
Figure 4.5 also shows that both unstable and stable air masses starve sheltered leeward slopes of moisture. This produces dry areas called **rain shadow** regions well beyond the uplands themselves. Once again, classic examples are found in the American deserts east of the Sierra Nevada (see Photo. 12). In Great Britain, for example, the Marchlands and Midlands leeward of the more modest 1000 m uplands of Wales are well known rain shadow regions. Similar effects are apparent across Eng-

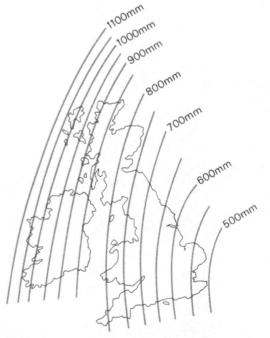

Figure 4.6   Annual rainfall over the British Isles assuming no relief (adapted from Bonacina 1945)

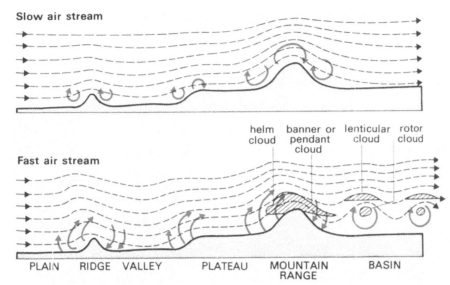

Slow air stream

Fast air stream

helm cloud | banner or pendant cloud | lenticular cloud | rotor cloud

PLAIN RIDGE VALLEY PLATEAU MOUNTAIN RANGE BASIN

Figure 4.7 Air streams over relief barriers

land with hills barely a third of this height. In fact, with prevailing winds off the Atlantic, the relief of Great Britain with uplands to the north and west causes most of the lowlands to the south and east to be rain shadow regions.

Figure 4.6 shows an attempt to estimate annual rainfall over the British Isles assuming a flat landscape. By comparing its isohyets (see Table 2.2) with *actual* ones on rainfall maps to be found in most atlases, a guide to rainfall amounts 'contributed' by relief can be obtained. The difference is sometimes called the **orographic component** of precipitation. For instance, it can be seen from Figure 4.6 that such places as Weymouth, Bristol, Manchester and Newcastle would all receive about 700 mm of rainfall each year if the country were flat. In fact, Weymouth gets about 50 mm more than this, Bristol about 100 mm more, Manchester over 150 mm more and Newcastle around 50 mm less. Thus, the northern Pennines create a rain shadow in the North East, but all the other examples given have extra orographic falls.

Precipitation is sometimes classified into orographic (or relief), frontal and convectional types depending upon the main processes producing the rain or snow. Total falls over intervals of time and even single storm events have been divided into the three categories by calculating the contributions made by each precipitation type. Suffice to stress here that it is easy to develop a fruitless circular argument as to their relative importance, short of careful measurements and mathematical treatments referred to as 'filtering' methods. This

is particularly the case with orographic and frontal rainfall which are so closely tied together in the British Isles, for example. It is best to regard frontal rainfall as a large-scale process and orographic contributions as modifications on a local scale.

The extent to which uplands affect the weather elsewhere varies considerably. Much seems to depend upon the movement of the air stream in relation to the size of barriers crossed. Like hurdling and high jumping, a good deal hinges upon the direction and speed of approach. Figure 4.7 gives examples of the currents resulting from slow and fast laminar air streams crossing various types of relief barriers. The red arrows show local eddy currents or **turbulence.** Generally, updraughts occur on windward slopes and downdraughts to leeward. With fast flows, especially, standing waves develop downwind of big hills and mountain ranges analogous to high seas breaking over a reef (see Photo. 13). Any clouds over local peaks will be streamed out like a banner with their tails pendant according to the strength of the leeward trough and downdraughts. The following wave crests may be identified by rolled or rotor clouds with lenticular cumulus above. The latter resemble an aircraft's wing section or aerofoil (see Photo. 14). Repeating patterns may exist up to 100 km beyond even modest uplands given strong air streams like the wake behind a ship.

Studies of air currents surmounting much smaller walls, fences, hedgerows and different types of tree show broadly similar patterns to those indi-

Photograph 13 Banded standing wave clouds. Strong general air waves passing over narrow mountain ranges in the Picos de Europa, Spain, form shallow cloud band along their moist crests.

Photograph 14 Lenticular leewave clouds. Strong local air waves downwind of mountain peaks in the Pyrenees produce lens shaped clouds (see Fig. 4.7).

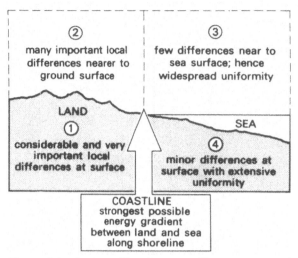

Figure 4.9 Orders of local mass and heat energy variations

cated in Figure 4.7. However, since winds can permeate through many of these minor relief barriers, the filtered air streams are retarded rather than disturbed into eddies. Such effects become apparent when snow drifts to leeward of shelter belts (see Photos 15 and 16).

In the same way that rubbing hands together creates friction, so layers of air blowing over each other generate forces giving rise to turbulence called **wind shear** (see p. 24). Figure 4.8 illustrates two forms of wind shear, one mechanical and the other thermal owing to a temperature inversion. Both produce rolling eddies in thin layers which become exaggerated when they occur together. Such turbulence characterises frontal surfaces and the upper air streams at the tropopause inversion. Banded or patterned cloud banks give visual evidence of wind shear (see Photos 4 and 5) but, more commonly, it produces unseen **clear air turbulence.** The latter, aptly called **CAT,** creates problems in flying, particularly at the roof of the troposphere where most modern jet aircraft cruise. High mountains below make matters worse.

We spend so much time seeking shelter from winds and draughts that they become an accepted part of our lives. Their importance in controlling local transfers of moisture and heat energy are considerable. Over hills and mountains these processes are heightened too and their effects more extreme and obvious.

C Processes along shorelines and coasts

In Britain, for example, nowhere is far from the sea by comparison with continental countries. Island weather is largely **maritime** and it is necessary to travel well into the heartlands of the continents before weather becomes. truly **continental.** The seas are an unlimited source of water vapour and they moderate extreme temperatures. Near the sea, excesses of moisture are countered by limited amounts of heat. All shorelines and coasts are the 'front lines' in the transfers of energy between air above the sea and land. Figure 4.9 illustrates this point with respect to the contrasting properties of rocks, soils and water.

Maritime effects
In general, two situations can be envisaged, one where the sea is warmer than the land and the other vice versa. Along some coasts where warm ocean currents penetrate into high latitudes, this

Figure 4.8 Mechanical and thermal wind shear

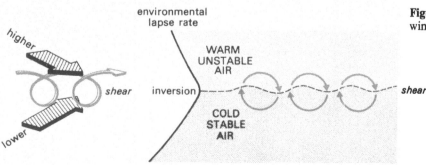

may be almost a permanent condition but, with most, such differences are either seasonal or diurnal. Each one is characterised by particular processes and distinctive coastal weather.

A given mass of water needs three thousand times the heat energy required by the same mass of rock to be warmed through the same rise in temperature. Furthermore, the rates at which water expands and contracts are much lower than in air. Water is not a compressible substance like air. In consequence, water masses are even much more conservative than air (see p. 48) with a great capacity to retain heat energy despite temperature variations overhead. Thus, large bodies of water temper extremes of warmth and cold on adjacent lands. Because mid-latitude westerlies dominate European weather, the effects of the Atlantic ocean remain dominant over continental influences as far inland as the Ural Mountains in the USSR. Using similar criteria, continentality is achieved more quickly in North America at these latitudes. Owing to the high mountain barrier along the Pacific coastline, which lifts the maritime air streams sufficiently to cause high orographic precipitation and modification, oceanic weather is confined to the westerly slopes facing the sea.

With no effective relief, water surfaces have widespread uniformity of temperatures reducing the possibility of local convection in the air over the open sea. Even within the water itself, convection is minimal and all large bodies have marked thermal stratification or layering. It is hardly surprising that the oceans make ideal breeding grounds for the bigger air masses. Many travel vast distances across surfaces with virtually uniform temperatures. Figure 4.10 shows that air masses passing over warmer currents are warmed from below so that their lapse rates become more unstable, e.g. polar air passing over the Gulf Stream

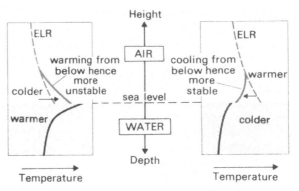

Figure 4.10 Water warming and cooling air masses

and North Altantic Drift en route for Europe. Greater stability occurs when warm air passes over cold currents, e.g. tropical air travelling over the Canary current towards the Mediterranean lands.

Fresh water has the unique characteristic of normally contracting and becoming denser as it cools down to 4°C; but, thereafter, it *expands* with further cooling. Thus, river and melt water is at its heaviest when at 4°C and will sink even through colder layers below. Because of the salts held in solution, however, sea water behaves differently and is heaviest just before freezing at about −2°C. Around the margins of the ice caps in the polar seas, therefore, outflowing summer melt water readily sinks and so does salt water later in the year as surface temperatures fall until sea ice forms. For most of the year, cold-water 'pools' are sinking in the polar seas and subsiding into the deeper ocean basins at lower latitudes. This is replaced by such warm surface currents as the Gulf Stream and North Atlantic Drift which carry thermal characteristics acquired in the Tropics well beyond the latitudes that air alone could achieve. Remember that ocean currents are responsible for transferring some surplus energy out of the tropics (see Fig. 3.15). It is this northward penetration of the conservative Gulf Stream and North Atlantic Drift, for example, that sustains the temperatures of tropical air masses en route to north-western Europe. Without them, Britain would be much drier but colder. Winter temperatures would drop by about 15°C if the Gulf Stream and North Atlantic Drift failed! Along the Norwegian coast it would be a devastating 25°C lower. Careful checks on both currents in the Atlantic show that their long term behaviour is crucial to the weather.

Salinity and surface winds exert greater controls over oceanic circulations in warmer waters

Photograph 15 Winds over walls. Snow drifts on the leeward side of a low wall indicate the shelter it affords. Deeper and longer drifts occur downwind of the highest sections of wall. In the open fields most of the snow has been blown away. The wind has blown from left to right of the picture.

Photograph 16 Winds through hedges. Snow drifts in a narrow country lane show how winds filter through hedges to deposit deep accumulations where velocities are least. In adjacent fields, the snow cover was considerably thinner. The wind has blown from left to right of the picture.

because thermal density differences are small. At the surface, salinity is largely a function of precipitation and evaporation. High rainfall and humidity in seas near the Equator reduce surface salinity to about 34 parts per 1000. In the mid-latitude anticyclonic regions, on the other hand, low rainfall and high evaporation increase open ocean salinity values to a maximum of 37 parts per 1000. Sea level in less-dense equatorial waters stands relatively higher than in the more briny water of mid-latitudes. Thus, wind-assisted currents flow literally downhill from the Equator polewards around the western margins of the oceans. Reference to an atlas map of ocean currents will show that this circuit or **gyre** is replaced by *upwelling* cold currents running along the eastern margins of the oceans from mid-latitudes towards the Equator. They are weaker and wider flows than the warm ones to the west, but they exert as great an influence upon the weather of the adjacent land masses (see Ch. 5.A).

Coastal weather

We may concentrate upon fogs, land- and sea breezes, storms and offshore currents as typical features of coastal weather. Figure 4.11 shows how **sea fogs** creep onshore during the afternoon and then lift as a layer of stratus cloud about 300 m thick at night when cold currents lie offshore. Notice the temperature inversion level which holds these local circulations down beneath a lid of warm air. Day time convection of dry air off the heated land surface sucks in cold moist air over the sea which forms an advection fog along the coast.

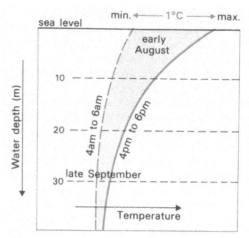

Figure 4.12 Variations of sea temperatures

At night, this is dispersed by the warmed dry air subsiding seawards which then rises and condenses beneath the inversion. Examples occur along the coastline of California (see Photo. 17) causing the famous sea fogs around the Coastal Ranges and Golden Gate, San Francisco. Similar conditions are common off the desert coastline of Peru and the southern coast of Portugal because of cold ocean currents offshore.

The conservative behaviour of water is also well seen on Figure 4.12. This illustrates the slight diurnal warming and cooling of water with depth typical of British seas. The warmest times of day and year are indicated, both getting later and less marked with depth. Note that late afternoon in early August is generally better for sea swimming at most European resorts, for example, although this is a couple of hours after the warmest part of the day and about a fortnight later than the hottest time of year. Throughout most of the day, the sea is relatively cooler than the adjacent land but at night it is comparatively warmer. Such diurnal reversals set up **land and sea breezes** which are particularly evident in summer. Figure 4.13 shows that these are effectively local convection cells (see Fig. 2.6).

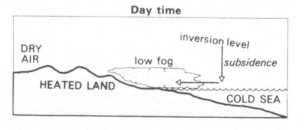

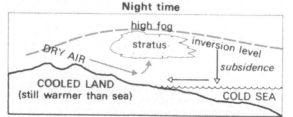

Figure 4.11 Coastal fogs and clouds with cold seas

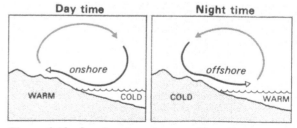

Figure 4.13 Land and sea breezes diurnally

Photograph 17 Sea fogs creeping inland. Afternoon sea mists thicken over a cold current, drift inland onto forested uplands and 'steam' off the warmer surface (see Fig. 4.11). (Taken in the Coastal Ranges of California near San Francisco)

With evaporation off seas, latent heat energy is withdrawn into the air. A cold film is established over the water surface, particularly when evaporative cooling reaches a peak. Condensation can even set in to produce a shallow **steam fog** covering the water surface. Sometimes, if offshore winds are very dry and hot, rates of evaporation are so great that sensible heat energy is taken from the air as well to support the mass transfers. This noticeably increases cooling along the shoreline. Because such processes are evident around large water surfaces in deserts, they are called **oasis effects.**

Calm 'flat' seas help evaporation. As winds get up, however, the rates slow down. The stable cold film breaks down over choppy water and the stirring of the lowest air layers reduces temperature gradients. Thus, sea to air energy transfers are less except where high winds cause spindrift, surf and spray. These can form fine salty mists that penetrate far inland during gales and storms.

Since most waves are wind driven, their erosive or constructive power geomorphologically is indirectly the result of the shoreline's exposure to weather at sea, especially prevailing winds. Blow hard across the surface of water in a bowl, and riffle or ripple waves form. These fan out downwind into distinct regular swell-like waves. Much the same happens at sea with the high winds that build up **storm** waves. Figure 4.14 shows how a

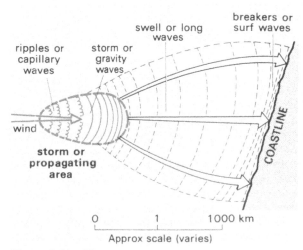

Figure 4.14 How storms start waves at sea

limited storm cell takes a hold of the water and propagates waves. These extend far beyond the confines of the storm. For a time, waves can even out-run the winds that drive them. If there is a net movement of water onshore, as around the British coasts, then corresponding sinking and submarine currents flow seawards. This causes dangerous bathing conditions on many beaches exposed to the 'open' sea or ocean, even if there is fine weather locally onshore. Conversely, with prevailing winds offshore, the net outflow of surface water is compensated by cold upwellings along the coast as mentioned already.

It can be shown that fluctuations in air pressure alone are only capable of depressing or lifting sea levels about 0·5 m either way in high- and low-pressure systems respectively. However, because such systems move and are large, their effects upon sea levels can be exaggerated. Just as dull sounds can set up vibrations in certain rooms, at particular speeds related to water depths, depressions strike up a **resonance** which lifts water surfaces up to 2 m above normal. A lot depends upon the direction of movement in relation to coastline configuration and tides. Such abnormal variations are called **seiches.** In large and busy estuaries such as Southampton Water, correction tables for normal tides are published with respect to barometric pressure, wind speed and direction.

When all factors work together in the open sea, waves between 10 and 30 m high can form. Fortunately, these moderate towards land; yet, **storm surges** are all too common in hurricane-prone tropical regions and they are not unknown around British shores. A well documented example occurred along the East Anglian coast when a deep low accompanied spring tides running southwards in the North Sea at the end of January in 1953. A resonance set-up which raised tides a full 2 m *above* predicted high-water mark in places. Severe flooding occurred, particularly in the Thames estuary. Not surprisingly, the Dutch have paid considerable attention to the accurate forecasting of such surges in the design of polder schemes.

Figure 4.15 provides a fitting conclusion to this section with respect to **offshore currents** along common types of coastline featuring submerged sills, e.g. enclosed seas, fjords, lagoons, barrier reefs and so on. Here, the limitations imposed upon the hydrological cycle both geomorphologically and climatically are important. In the evaporative situation there is a net loss of mass, but a gain of heat energy leading to stagnant pools of

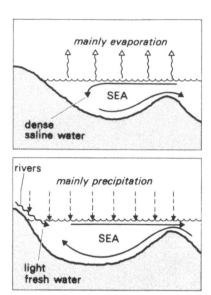

Figure 4.15 Circulations in silled seas with dry and wet climates

warm saline water deep in the basin. These build up to kill off marine life there; the Black Sea and Mediterranean Sea are cases in point. On the other hand, the precipitation-dominated régime is more balanced and full of life. Similar situations have been observed in large lakes and manmade reservoirs in delicate energy-balance terms. Local processes in the air are vital to life in the water too!

It is too easy to forget that 70% of the air's lower boundary is in contact with ocean surfaces (see Fig. 1.2). Life-giving moisture in the air relies upon this large area of contact because, even though it reaches almost everywhere, water moves quickly through the atmospheric system. Indeed, if all the precipitable water held in the air were to be suddenly released, it would represent a downpour lasting only a few hours.

D   Processes on plains and slopes

This section starts literally in lighter vein by taking up the story on the beach and then moving further inland.

Exposure to sunshine

It is well known that a suntan is acquired more readily by the sea because of the greater scope for the reflection of shortwave light off the surf and sand surfaces. The sea itself only reflects sunlight well when this shines at low angles across the water in the morning and evening. Since morning air is

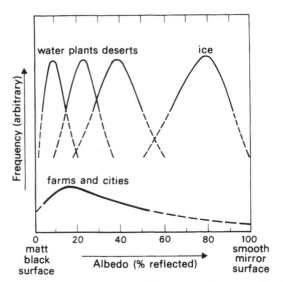

Figure 4.16 Albedo of different surfaces to all radiation

invariably clearer, beaches facing east and southeast tend to record more hours of bright sunshine. Cliff vegetation often flourishes in response to morning light and different species favour slopes facing various directions and angles regarding both direct and 'bounced' radiation.

Think of setting up a deck chair and the three key factors involved in one's subsequent comfort. They will be the albedo (see Ch.3.D) of the clothes or skin, the **aspect** (direction) faced and the **angle** (slope) reclined with respect to the position of the Sun in the sky. Since all land surfaces comprise slopes with these characteristics, much the same applies to them as well. Even plains are slopes with little or no angle!

Figure 4.16 indicates the albedo of different surfaces using arbitary frequency curves. Ice has the highest **albedo** but different types of snow surface will cause such values to vary quite a lot. Water, on the other hand, has a low albedo and little variation. Remember that it covers most of the Earth's surface, however, which is why average global values are about 30%. Notice that most manmade landscapes are less than half as reflective, although considerable variation does exist. On the whole, Man's activity seeks to increase the amount of solar energy absorbed and used. We see some of the consequences in the next section. Suffice to point out here that direct radiation reflected off a surface remains unaltered so that the greater proportion, if not all, returns unused into space (see p. 47). This explains why snow

covers survive sunny weather even after air temperatures have risen above freezing point: dirty snow melts more quickly.

Imagine that you are on a plain and lying flat on your back with feet pointing due south. The view of the sky appears as a dome overhead. The *horizon* of the global surface is in the plane of the body, and the line of sight vertically upwards is known as the **zenith.** By lying still all day, you would see the Sun appearing to travel across the dome of the sky from east to west. Its path is a function of both the *latitude* of your position and the *day* of the year, because of the situation already described in Chapter 3 (see Fig. 3.14). Should you remain prone for a whole year, each day's path would be slightly different, lowest across the sky at the winter solstice and highest during the summer solstice. On both equinoxes, the Sun would rise due east, set in the west and attain a noon angle due south from the zenith. This angle would be the same as the latitude of the place.

Figure 4.17 is an accurate drawing of the Sun's paths across the sky over Wells, Somerset, at latitude 51°N and longitude 2½°W. Other localities have slightly different paths across such models, as will be seen in the last chapter (Figs 5.1 and 5.3). Concentric rings show angles of elevation from zero at the horizon to 90° at the zenith in the centre. The radial lines from this centre are the points of the compass as bearings. Sun paths are plotted for both solstices and equinoxes and then divided by gentle time curves. Note that **local noon** is 10 minutes after midday Greenwich Mean Time (GMT) because Wells, Somerset, lies 2½°W of the **Standard Time** meridian (longitude 0°) for the British Isles.

In much the same way that a nearby object or person on the plain could screen you from direct sunshine when you were lying down, so features above the horizon might blot out the solar beam and cast shadows. When shaded, a site becomes reliant upon diffuse or counter radiation for receipts of heat energy (see Ch. 3.D). Such indirect warming is usually weaker. However, if the screen is elevated to more than 30° above the horizon, even diffuse sky radiation is less effective. This happens in woodland clearings, in deeply incised valleys and, of course, along streets with tall buildings. Figure 4.18 uses a simplified Sun path model for Wells, Somerset, to show how a flat garden near the city is screened from direct sunshine. Notice how it will be in the shade of buildings and trees for about a month from mid-December to

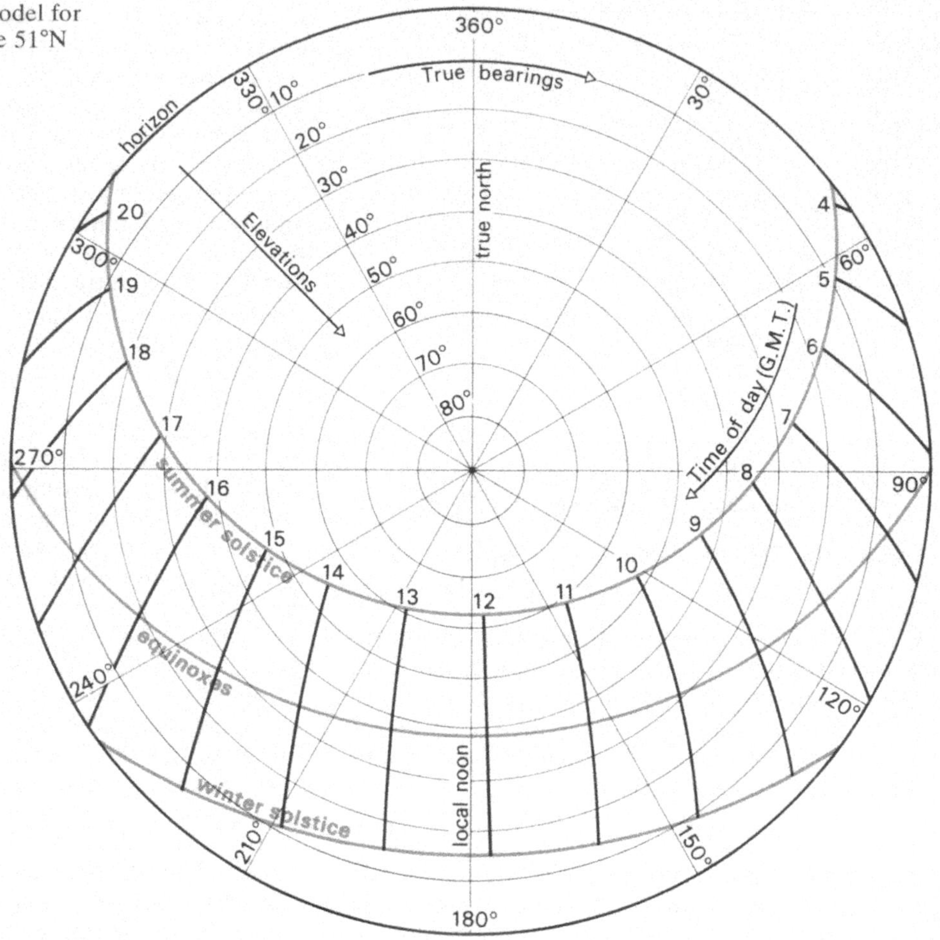

Figure 4.17 Sun path model for
Wells, Somerset, Latitude 51°N
Longitude 2½°W

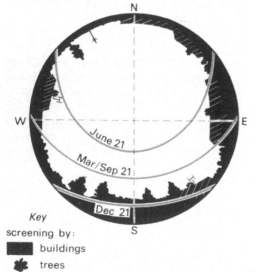

Figure 4.18 Screening of a garden in
relation to the Sun at Latitude 51°N

Key
screening by:
■ buildings
❀ trees

mid-January. Furthermore, it is more exposed to
colder northwesterly winds.

The warming of slopes
Increasing variations of slope aspect and angle
normally occur on moving inland. Both can be
plotted onto Sun path diagrams in much the way
that the deck chair's direction and inclination can
be altered. Figure 4.19 illustrates two contrasting
examples from the north and south-facing slopes
of steep hills in southern England. In each case, the
basic model is suitably truncated. It can be seen
that a steep slope facing south-west deprives itself
of morning sunshine except during the winter solstice. On the other hand, less steep north-facing
slopes are screened and shaded for most of the
winter. To each might be added the effects of
buildings and trees, of course, as in Figure 4.18.
In small areas at higher latitudes, different slope

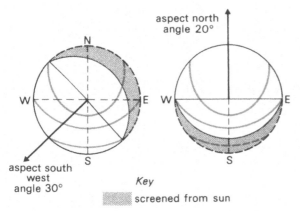

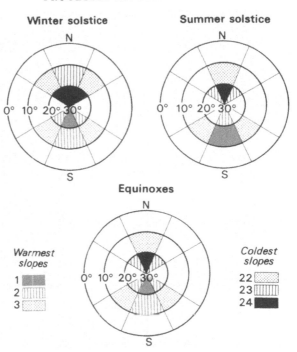

Figure 4.19 Effects of slope aspect and angle upon screening out the Sun

Figure 4.20 Warmest and coldest slopes on hills in southern England with angles up to 30°

aspects and angles give rise to contrasting local heat-energy budgets closely related to amounts of direct solar radiation received and absorbed. If we imagine a conical hill with concave slopes reaching a maximum of 30° at the summit, it might be shown as one of the drawings on Figure 4.20. Aspects are given by the eight radial wedges, and angles by the concentric rings. These produce a total of twenty-four slopes to be warmed and cooled. The three diagrams on Figure 4.20 show how the warmest and coldest slopes occur at the main seasons of the year, all other things being equal. Notice how the warmest slopes are less steep ones in summer and that the coldest areas shrink around the north-facing shoulder.

Temperature measurements on such slopes reveal that the greatest contrasts are among the steeper slopes during spring and autumn. The smallest differences occur during the summer, especially on gentler slopes. These characteristics are vital regarding plant growth; for example, steep south-facing slopes can be up to a month 'earlier' in producing crops of spring vegetables and fruit. There is a tendency for diffuse sky radiation to cancel out such differences on low-angle slopes but to exaggerate them once above 30° in areas of strong relief. Buildings are cases in point with vertical walls facing different directions (see Ch. 4.E).

Now, imagine the slopes in Figure 4.20 to be reversed so that they show conical basins rather than hills. Diffuse sky radiation becomes even more effective in sharing out heat energy between different slopes and, of course, ventilation is reduced owing to shelter from the wind. In consequence, basins are invariably warmer than corres-

ponding hills with similar slopes. This makes them ideal 'hot spots' inland above which convection currents develop particularly in summer (see Ch. 2.D). The incidence of convectional showers and late afternoon thunderstorms is much greater over interior basins, e.g. the Midland and London basins in England. Such convectional rainfall is also characteristic of warmer continental interiors during summer (see p. 67).

Figure 4.21 shows three stages in the life of a convectional thunderstorm. Given that it is fed by enough moist air, deep cumulo-nimbus clouds form over 'hot spots'. Extreme instability (see Ch. 3.E) can lead to updraughts or thermals over 30 metres per second. Turrets of water droplets unfold at the edges of the cloud and draw in surrounding air. This process of cloud growth has been called **fibrillation.** Bergeron processes (see p. 53) occur in the freezing layers and there is an intermediate stage when large ice particles fall through water droplets still rising. Finally, the mass of precipitating water takes over and the cloud collapses, driving chilling winds ahead of the downpour.

Thunder and lightning often accompany convectional storms because the ice particles aloft hold positive electrical charges (see Ch. 3.B and 3.E).

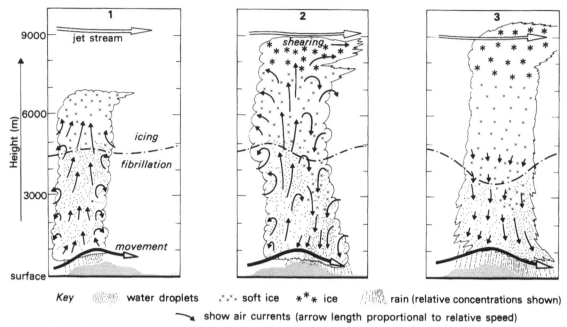

Key [water droplets] [soft ice] [ice] [rain (relative concentrations shown)]
[show air currents (arrow length proportional to relative speed)]

Figure 4.21 Heavy convectional storms

These build up to a high potential then suddenly discharge to Earth through the cloud. At much smaller scales, similar effects can be felt with static electricity on the ground. Since the channel of a lightning stroke instantly heats to an incredible 15 000°C, explosive expansion of air creates the thunderclap. Persistent lightning, therefore, assists turbulence within thunderclouds.

Heated inland basins in South-West England, especially, can cause heavy downpours in summer when fed with deep moist tropical air. Figure 4.22 shows a rainfall map for such a storm crossing the narrow Mendip Hills, England, from the lowlands of central Somerset into the Avon basin near Bristol on the night of Wednesday 10 July 1968. Notice how the heaviest falls were *leeward* of the

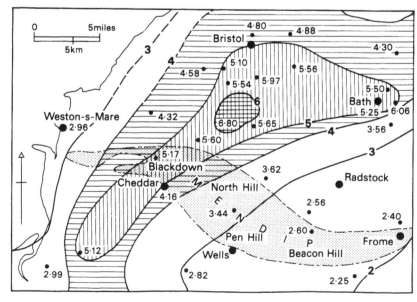

Figure 4.22 A heavy rainstorm (rain gauge measurements taken in inches)

Figure 4.23 A rainstorm surface

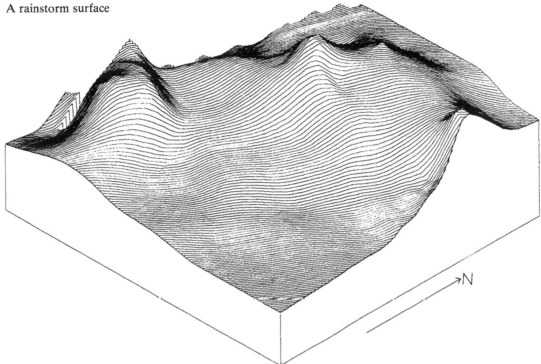

hills in what would normally be a rain shadow area with orographic precipitation. Valleys can even exert controls on the route followed by thunderstorms. Figure 4.23 shows the depths of rain that fell in this storm as a continuous surface looking north west from a vantage point above Wells. Somerset. All storms have similar 'hat-like' surfaces. These will often be a mirror image of the relief below, i.e. a hat on a basin. Fortunately, violent storms like this do not return often to the same place and their occurrence is unreliable.

Although we are now inland and far from the beach, it is worth remembering that the heat energy alone off warmed slopes could not create convectional storms without the addition of moist air from the sea. It is the mix of both that is important.

E Processes above urban areas

In this concluding section we move further into the realm of mixed processes and see that particular *patterns* or recipes apply to major urban regions. Since most people live in or near large settlements, we also have an opportunity to bring together the main processes outlined so far within the context of familiar everyday environments.

Heat islands

Actually, of course, there are no individual atmospheric processes as such peculiar to the air above urban areas. Rather, we see Man's activities subtly exaggerating some processes and suppressing others. By grouping urban activities and concentrating higher energy consumption, city air becomes warmed as a **heat island.**

When processes alter in the natural environment they have a happy knack of being returned into balance by equal and opposite reactions elsewhere. This is called **negative feedback.** Growing concentrations of human activities in towns, however, seem to have opposite effects called **positive feedback.** Here, natural balances are upset and a change in one leads to appropriate changes in all. This is hardly surprising since we have seen already that Man seeks to get more out of the atmosphere than nature in certain places. Cities are regions where these demands are most evident.

Research into modifications of the natural blend of processes between town and country has focused upon single phenomena to date. A profile of the whole impact of urban development is only just emerging. Most work has been done in major industrial regions of Europe, America and, more recently, in Japan. Two discoveries by climatologists working in London can be highlighted; Professor T. J. Chandler has demonstrated the heat island effect of central London, along with measurements of airborne dust concentrations, and Dr B. W. Atkinson has shown that increased convectional rainfall occurs during summer. There is a clear positive cause and effect link between both findings.

The burning of fossil fuels releases additional heat energy and gases such as carbon dioxide and water vapour which absorb outgoing radiation. Dust provides hygroscopic nuclei to increase the incidence and intensity of fogs (see p. 41). In heavily polluted air, the combination of smoke and fog produces **smogs.** Both dust and fogs reduce visibility and reflect a lot of direct incoming radiation. Thus, radiation exchanges are reckoned to exert the most important controls over urban heat islands.

The production of radiant and convective heat energy in some cities can match and exceed that received at certain times of year. Values for the former are largely estimates, of course, and measurements of incoming radiation usually represent *totals* of direct, diffuse and reflected components to *horizontal* surfaces *averaged* out over given times like a day (24 hours) not taking the daylight hours into account. In many big north European cities, mean outputs of energy from heating systems during winter may reach 25 W m^{-2} to exceed that coming in from the Sun to *horizontal* surfaces on overcast days. If allowances are made for the *vertical* surfaces that dominate tall buildings and their consequent shading, then such outputs are likely to be much greater than inputs. Even during clear summer days in heavily industrialised centres, the production of energy by manufacturing processes can approach 10% of that coming in, i.e. about 250 W m^{-2}, to exposed horizontal surfaces. In the manufacturing regions of the northeast of the United States, for example, mean radiation inputs to flat surfaces are about 60 W m^{-2} in mid-winter rising to over 260 W m^{-2} in mid-summer. Comparable data for cities like Las Vegas in the desert state of Nevada vary from a pleasant 120 W m^{-2} to a blistering 360 W m^{-2}. Here it must be remembered that air conditioning plants add to the heat-island effects. Owing to the clear air, desert surfaces collect more radiation than do those in cloudy equatorial regions. On average, places near the Equator receive between 150 and 200 W m^{-2} with little variation.

Ventilation is of secondary importance in regulating the heating and cooling processes. City centres up to 2°C warmer than the surrounding countryside are evident during calm and stable conditions, but turbulent air blows away the dust and fumes responsible for warming. In these circumstances, the effects of the heat island are either blown downwind of the city or are largely dispersed when wind speeds exceed 5 metres per second (18 km h^{-1}). The so-called **mixing length** of smoke plumes is a function of wind speed and air turbulence, both of which are strongly influenced by the size and shape of buildings. In fact, the 'roughness' of the built-up area seems to be an important feature with respect to ventilation and mixing.

Table 4.1 summarises the conclusions of most research on urban climates to date. Modifications to processes involving the six main elements of weather (see Table 2.2) are seen to result from the changing composition of the *air* and the surface of the *ground*. By referring back to the appropriate explanations given in the last two chapters in particular, Table 4.1 forms a fitting way of revising the processes outlined so far.

A fuller appreciation of atmospheric processes is clearly required of architects and urban planners as the work by Lacy for the Building Research Establishment in Britain shows (1977). During the Industrial Revolution in Britain, for instance, the less polluted areas were usually to the west side of the factories because of prevailing westerly winds. Such upwind locations became more fashionable, and this strongly affected the contemporaneous development of residential suburbs and manufacturing districts. Although there are many other factors to reckon with in town planning today, there is a reawakening to the positive benefits of materials and designs which take advantage of the *local* climate. Good examples are the orientation and surface areas of walls and windows exposed, the shapes and layouts of buildings to make the best of natural warming and ventilation, and so on. In particular, the virtues of improved home insulation – rather than more costly heating or air conditioning – are now widely recognised as being economically preferable as well as environmentally sound. Factories and homes are no less influenced by the weather than are farms and gardens.

Global warming
There is now little doubt that the whole atmosphere is warmer as well as the air over cities. Mean temperatures at the surface have risen significantly

Table 4.1 The characteristic mix of atmospheric processes over urban areas

Structure of the air (see Ch. 3.B)

Greater amounts of dust increasing concentrations of hygroscopic particles. Less water vapour but more CO_2 and higher proportions of noxious fumes owing to combustion of imported fuels. Discharge of waste gases by industry.

Structure of the ground (see Ch. 4.D)

More heat-retaining materials with lower albedo and better radiation absorbing properties. Rougher surfaces with vast variety of perpendicular slopes facing different aspects. Tall buildings very exposed and deep streets sheltered and shaded.

Resultant processes (see Table 2.2)

1. Radiation and sunshine (see Chs 3.D & 4.D)
Greater scattering of shorter-wave radiation by dust but much higher absorption of longer waves owing to surfaces and CO_2. Hence, more diffuse sky radiation with considerable local contrasts owing to variable screening by tall buildings in shaded narrow streets. Reduced visibility arising from industrial haze.

2. Clouds and fogs (see Chs 3.C & 4.A)
Higher incidence of thicker cloud covers in summer and radiation fogs or smogs in winter because of increased convection and air pollution respectively. Concentrations of hygroscopic particles accelerate the onset of condensation (see 5 below).

3. Temperatures (see Chs 3.E & 4.D)
Stronger heat energy retention and release, including fuel combustion, gives significant temperature increases from suburbs into the centre of built-up areas creating heat 'islands'. These can be up to 8°C warmer during winter nights. Heating from below increases air mass instability overhead, notably during summer afternoons and evenings. Big local contrasts between sunny and shaded surfaces, especially in the spring.

4. Pressure and winds (see Chs 3.A & 4.B)
Severe gusting and turbulence around tall buildings causing strong local pressure gradients from windward to leeward walls. Deep narrow streets much calmer unless aligned with prevailing winds to funnel flows along them.

5. Humidity (see Chs 3.C & 4.A)
Decreases in relative humidity occur into inner cities owing to lack of available moisture and higher temperatures there. Partly countered in very cold stable conditions by early onset of condensation in low-lying districts and industrial zones (see 2 above.).

6. Precipitation (see Chs 3.C & 4.D)
Perceptibly more intense storms particularly during hot summer evenings and nights owing to greater instability and stronger convection above built-up areas. Probably higher incidence of thunder in appropriate locations. Less snowfall and briefer covers even when uncleared.

by as much as 1·0°C almost everywhere during the past century. The extent to which this can be attributed to natural or human processes is still debatable, however. Suffice that the consensus of research strongly suggests how faster rates of industrialisation and deforestation worldwide are largely to blame.

It is estimated that we dump about 6×10^9 tonnes of carbon dioxide into the atmosphere every year by burning fossil fuels and a further 2×10^9 tonnes through farming. Not all of this can be recycled and, so, CO_2 concentrations must increase. Measurements show average increases from 280 parts per million to the present global mean of 315 ppm in the past 100 years. Values of about 350 ppm have been predicted by the year AD 2000. Since the amounts of CO_2 set the temperature of the air (see Ch 3.D), the worst estimates indicate further warming by as much as 0·5°C.

In addition to the oceans, which exchange something like 100×10^9 tonnes of CO_2 with the atmosphere annually, living organisms cycle 60×10^9

tonnes per annum. Forests account for most of the latter and these are being cleared rapidly. Unless the oceans can cope by storing more, preferably in new sedimentary rocks, there is no natural agency preventing the build-up of CO_2 in the air. These increases are regarded as much more ominous than the depletion of oxygen through deforestation. Although the world's forests produce about 55×10^9 tonnes of oxygen annually, over a quarter of which comes from the tropical rain forests, this amount is tiny by comparison with the total available throughout the atmosphere.

We must conclude that the links between deforestation and the global warming of climates are indirect. Broadly, the increasing world population with relatively more people living in towns demands greater productivity from dwindling farmland. By clearing forests, tilling soils and then growing crops or grazing animals, we remove the canopy intercepting radiation and precipitation, disturb the soil and concentrate fewer species into smaller areas. In primary forests, different species are widely scattered. On farms they are brought together to experience greater variations of heating and cooling along with extremes of wetting and drying. Furthermore, the rich vertical layering of environments is drastically reduced when tall trees are felled and replaced by short crops. The resultant horizontal and vertical concentrations of organisms lead to increasing pressures upon available foods in a given space. Gradually, producers cannot meet the demands of the consumers, and scarcities occur. Once such deficiencies take hold, extremes of weather can lead to irreversible soil erosion by too much water or heat locally.

In substituting for the shelter of the forest that afforded by the city, Man exchanges an efficient and stable system for a less efficient and apparently unstable one. The forest has a low nutrient store rapidly cycled by *moisture,* but the city requires larger food stores which are produced mainly by *heat* energy. In effect, the former is a self-regulating water-cooled system whereas the latter has a tendency to overheat.

Other mechanisms which contribute to changing global climates are mentioned at the end of this book (see Ch. 5.C). We may conclude here by confessing again that urbanisation might well be *reinforcing* the natural processes causing changing climates at the present time or, equally unwittingly, be *countering* them. There are supporters for both viewpoints, although most evidence currently favours the theory that bigger towns and more industrial activity are major reasons for the global warming of the atmosphere in recent times. The best indication of this trend is probably the greater warming over the industrialised temperate latitudes of the northern hemisphere by comparison with other climatic zones.

Chapter 5

Global Patterns

Having accounted for the key processes in the atmosphere at all scales and touched upon how they blend together in the more familiar setting of towns, it remains to survey the global scene likewise. From such a base, world climates come alive and mean more than mere collections of descriptive statistics.

So, by now, the reader should feel better equipped to tackle the selected modern texts that are listed at the end of this book. As many of them analyse the *patterns* that distinguish zonal and regional blends of processes through case studies, they satisfy the traditional quest for information that supports the geographical description of climates. Here we must be content with concise recipes of what to look for and to expect as our emphasis has been given to basic physical processes.

But, first, a word of warning! Irregularities in atmospheric processes are normal and not exceptional. Although our atmosphere is very old, we have had reliable instrumental records only for the past century and an adequate global network of stations for a fifth of that time (see p. 11). This is a very tiny sample and, since most of the climatologists' models of process patterns are derived and tested statistically, they are subject to all the usual hazards of statistical methods based upon scanty evidence. Each physical process can be reliably measured, but we must be rather more cautious when relating several together in studying the very complex patterns apparent in the atmosphere as a whole.

No climatologist would claim to have complete answers to the process – pattern relationships. This uncertainty is best explained by reference to the limitations of **climatic normals,** i.e. averages of temperature, rainfall and so on. The 1935 Congress of International Meteorological Organisations agreed upon the 30-year period 1901–1930 for the first set of normals. Since then, we have only completed one other, from 1931 to 1960. Significant differences exist between the two sets of normals which have led to teams investigating the strong probability that the global patterns are changing, e.g. Professor H. H. Lamb in Britain and Professor Reid Bryson in America.

It is a strict rule in statistical analysis that the time it takes for similar events to recur, called **return periods,** cannot be predicted *confidently* for intervals longer than the time over which the data has been collected. Thus, no one can be really sure about changes spanning more than a human lifetime let alone longer ones. Since future data cannot be invented, we must turn to the back log of historical and geological evidence for help, or await further advances in theory. The former has been favoured by Lamb and the latter by Bryson. Both are complementary.

In short, look further than the thumb nail sketches that follow, especially at the well documented case studies for particular areas present and past. See whether you can apply to them the principles learnt about atmospheric processes here.

A Patterns in low latitudes

For all practical purposes, the **tropical regions** start at latitude 40° when travelling towards the Equator. They occupy 64% of the Earth's surface area, added to which their weather patterns are wholly different to all other zones. For these reasons alone, the low latitudes deserve more attention. Remember, too, that the atmosphere boiler room lies here and that moisture is as important as heat energy. Indeed, tropical seasons are best identified by rainfall régimes rather than temperatures.

The atmosphere in the Tropics

The general circulation within 40° either side of the Equator is dominated by the Hadley cells (see Figs 2.9 and 2.28). At the time Sir George Hadley advanced his theory, sailors plying the infamous slave trade route to West Africa and the Caribbean recognised three belts of tropical weather: fine settled conditions from 40°N to the Tropic of Cancer at 23½°N, welcome prevailing north-easterlies to about 5°N, and then a narrow band of calm but invariably 'dull' weather near the Equator itself. These belts became known as the 'horse' latitudes, trade winds and doldrums respectively. Only the trade winds term survives to describe the surface wind belts immediately to the north and south of the Equator. Today, three similar divisions are still recognised, but they are better renamed as the **subtropical high pressure** regions or divergence zones, the **easterly air streams,** and the **intertropical convergence zone (ITCZ).** As the last two lie between the Tropics proper, they are sometimes classed as *intertropical* features as distinct from the *subtropics*.

Formerly it was thought that a continuous equatorial low pressure belt or trough ought to exist with a distinct intertropical front between the converging trades from either hemisphere. There is no actual evidence for this, however, and so the notion of the ITCZ is preferred. It may be defined as the poleward limit of the general meeting between the trades and it appears to form a discontinuous ribbon spanning about 5° of latitude itself. Within it lies the elusive **equatorial trough.** The ITCZ swings north and south of the Equator with the seasons, notably over the land. It coincides roughly with the heat equator (see Fig. 2.8) at the equinoxes and is less active at the solstices. Sideways or lateral movements are relatively easier in low latitudes because there is a much greater thickness of atmosphere above the Equator (see Fig. 2.28). So, there is more room for different air masses and streams to pass over and under each other to form contrasting vertical layers of air in the Tropics.

Because vigorous convectional overturning of air cells occurs within the ITCZ, latent heat is withdrawn from the tropical seas, soils and rain

forests to be released aloft when condensation sets in. Deep clouds form leading to tropical storms. So, the dried and warmed air near the tropopause is then driven north and south at the top of the Hadley cell circulations. This means that the surface convergence is balanced by an upper divergence.

Thus, warm dry air lies over a shallow surface layer of cooled moist air bordering the ITCZ. This gives rise to a marked inversion above the trade winds and subsiding warm air within the subtropical anticyclones (see Figs 2.10 and 2.23). Such layering accompanied by widespread uniformity of temperature and pressure effectively makes the tropical atmosphere barotropic (see Ch. 2.D). Hence, air masses in the mid-latitude sense do not exist and fronts cannot form. Tropical weather is viewed as different types of *disturbance* forming locally within major *air streams*. A paradoxical mix of regular local disturbances of a violent nature within slack or uncertain patterns is the lot of equatorial forecasters in particular. Satellite photographs are a boon.

Features of tropical weather

Several factors combine to explain the weather of the tropical world. Foremost is the weak Coriolis force (see Table 2.3) up to 10° of latitude either side of the Equator. So, although temperature and pressure differences are small, their tiny gradients rule and winds blow across isobars perpendicularly rather than revolving geostrophically as at higher latitudes (see Fig. 2.21). In short, there is no induced global spin or vorticity (see Ch. 2.G) near the Equator. Most **equatorial weather,** therefore, is a local matter related to latent heat energy transfers through evaporation and condensation (see Ch. 3.C). Only within the trade wind belts or tropical easterlies do there appear to be organised patterns of activity.

Figure 5.1 illustrates the main features involved in **tropical weather** further from the Equator. Note the important ocean gyres (see Ch. 4.C) and the 'spots' which indicate patchy upwellings of cold water in the oceanic 'watershed' along the Equator, yet another effect of the weak Coriolis force there. Figure 5.2 takes a closer look at the more active summer régime in the northern hemisphere. Its clear threefold pattern is summarised below.

Within the ITCZ, the disturbances are local, virtually stationary and non-revolving. They appear to be small-scale convectional 'pumping'

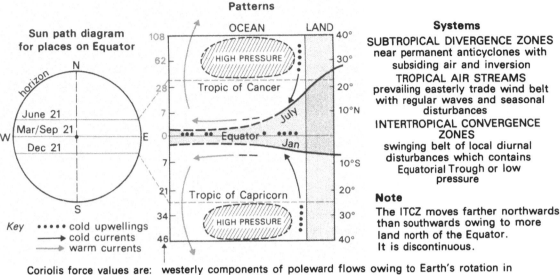

Figure 5.1   The world of tropical weather

associated with adjacent warm and cold patches of water or lands for that matter. Groups or clusters of cumulus clouds build up diurnally over hot spots. Afternoon and early evening thunder can be monotonously regular. Clear skies occur with air subsidence over colder places.

In the trade winds or easterly air stream further north (or south), a 'river' of moist air streams beneath the shallow inversion with dry air above. Owing to wind shear and curvature (see Fig. 4.8), its surface at the inversion becomes riffled and then waved. Like slowly moving waves in fast flowing rivers, streams of air move up and down

faster than the waves themselves are drifting. Such motions in the **wave trough** shown in Fig. 5.2 are zones of divergence and convergence (see Ch. 2.E). As the waves become bigger further west, they suck up moist air off the ocean. Because this has been warmed on its long sea passage, it is highly unstable. Therefore, once the trade wind inversion weakens westwards and its restraining lid of warm air is removed, unstable parcels burst aloft like jacks-in-the-boxes. The majority of tropical thunderstorms along the western shores of the oceans at the ends of the easterly air streams occur like this (see Photo. 6). The

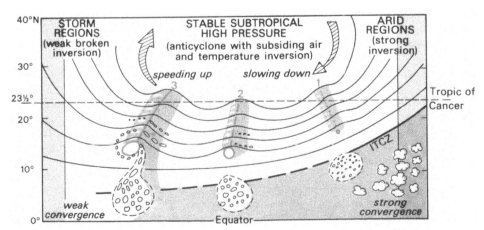

Figure 5.2   Summer in the tropics across the oceans between continental margins

maxim, 'A wave a week', applies well to this situation.

Hurricanes (in the Caribbean) and **typhoons** (in Asia) occur at the end of summer when the ocean beneath the trades is at its warmest. The hurricane season may be thought of as a period of rougher wave action, with such places as the Caribbean and the shores of South-East Asia taking the brunt of the onslaught. Just as one sea wave in a succession seems to be much bigger, so occasional troughs in the easterlies are large enough to draw in storms from the ITCZ, as is shown in Figure 5.2. The **eye** or column-like core of the system may begin above a cooler patch of sea in the trough of a bigger easterly air wave. Warm and moist air with storms is drawn towards the disturbance to form a thickening vertical wall of clouds like a revolving cylinder or sleeve around the calm and cooler core. The release of latent heat with so much resultant condensation appears to warm the core area suddenly by as much as 10°C. Therefore, warm dry air now subsides in the eye. Its adiabatic expansion and further warming at the surface makes it ripe to suck moisture off the surrounding ocean. In effect, the spinning system can now feed itself from within. As the system moves over warm seas, such tropical cyclones intensify and grow into hurricanes. When moisture supplies are cut off overland or, more rarely, over cold currents, the system quickly dies. Therefore all such tropical cyclones must be maritime. Also, beyond latitude 20°N they curl to the north-east and speed up with the Earth's rotation. The mirror-image pattern occurs south of the Equator. Being unable to feed adequately while moving faster, hurricanes fill and weaken like more modest extra-tropical lows.

Thirdly, the subtropical high pressure systems **(STHP)** are firmly moored to the cold updwellings and currents along the eastern shores of the oceans from about 30°N to 35°N. The very stable inversion overhead here is only 500 m high. Subsidence with adiabatic warming over adjacent hinterlands maintains clear skies and so the aridity characteristic of tropical deserts such as the Sahara. These dry zones are the roots of the trade winds which drive easterly air streams to the far side of the ocean and cause storms there. Clearly, vast quantities of water vapour are picked up by the trades on this journey, indicating a big shift of both mass and heat energy westwards in these latitudes. In addition they hold the 'stolen' momentum gained by moving against the Earth's rotation (see Ch. 2.F).

Distinctive climatic types result from the three patterns outlined. Respectively, these are the **equatorial** or **selva** climate with rain all the year, the **savanna** with a dry winter and wet summer and, of course, the **deserts**. **Monsoon** climates are an extreme variety of savanna régime to be found mainly in India and South-East Asia. Also, it is worth noting that the poleward margins of the deserts fringing the subtropical regions contain the **Mediterranean** climates with wet winters and dry summers. Thus, all seasonal climates within 40° of the Equator are dominated by varying *moisture* régimes as the belts of tropical easterlies swing north and south with the ITCZ. However, since the study of regional climates in depth is beyond the scope of this book, the characteristics of the types given are best examined through examples from the literature appended to this final chapter.

In regions where water cycling is so important, evidence for climatic change is most likely to be found in the oscillating shorelines of large freshwater lakes. These act like giant long-range rain gauges and evaporating pans. From Lake Victoria in East Africa, for example, there is good evidence of a 2 m rise in water level since 1960 following more rain to its large catchment area in the surrounding equatorial mountains. On the other hand, since 1968, rainfall has been much lower in zones between 14°N and 20°N along the savanna and desert margins. Failures of essential summer rains in the parched Sahel fringing the Sahara Desert suggest that the moist airstreams involved are less reliable or have ominously shifted southwards. Other evidence supports a zonal shift of the rain-bearing air streams towards the Equator. We will review likely causes in the last section of this chapter. In less than a decade, such changes have already proved socially and politically disastrous in these areas.

B Patterns at middle and high latitudes

Beyond latitude 40°, *temperature* increasingly takes over as the dominant feature of weather patterns. With shorter zonal circuits around the Earth, the Coriolis force becomes stronger too. Remember that considerable gains of both thermal and mechanical energy in the Tropics are transferred into higher latitudes. Consequently, although middle and high latitudes comprise barely more than a third of the Earth's surface area, the westward shifts of energy seen in the Tropics are balanced polewards by more vigorous transfers eastwards. More water is locked up as ice because seasonal temperatures hover around the critical threshold between freezing and melting. In many respects, the weather patterns are the reverse of the tropical world.

The atmosphere in polar and temperate regions
It is worth remembering (see Ch. 2.G) that the general circulations in extra-tropical regions are mainly the result of upper westerlies or Rossby waves meandering around the lower roof of the troposphere. The poleward margins of the westerlies contain the fickle **polar front jet streams** and the 'outer' fringe the steadier **subtropical jet stream** (see Figs 2.26, 2.27 and 2.28). High mountains and regional contrasts in the heating of land masses and ocean currents provide the necessary mechanical and thermal gradients to produce waves in these streams. They stir and mix the air below by driving chains of swirling depressions eastwards. Contrasting tropical and polar air masses drawn into these systems (see Ch. 3.E and 4.A), result in strongly baroclinic situations (see Ch. 2.D) with fronts. Again, this totally contrasts with the Tropics where most wave patterns were seen to work upwards from the surface under more barotropic conditions rather than downwards.

Sun path diagrams every 10° on a poleward march from latitude 60°N are given in Figure 5.3. Compare them with that for 51°N on Figure 4.17. At 60°N, the Sun at the winter solstice only climbs to a maximum noon angle of 6½° above the southern horizon. A comparison with Figure 4.18 will indicate that low mountains easily screen out direct solar radiation in winter at such latitudes. North of the Arctic Circle at 66½°N, the noon Sun at the winter solstice is 'lost' below the horizon. Refraction and reflection off clouds and snows give a twilight, however. Thus, the **polar** world of long winter nights balanced by perpetual daylight for increasing summer periods effectively begins at latitude 70°N. This means that the **temperate** zones lie between latitudes 40° and 70°. Notice, at the North Pole, how the Sun at the equinoxes runs around the horizon, climbs to the June solstice and sinks back again for a 'day' six months long. A frigid winter 'night' follows for the next six months.

Features of polar weather
Apart from the *durations* of radiation, the summer Sun tracks on Figure 5.3 show that slope angle and aspect (see Ch. 4.D) become less important polewards since direct sunlight comes from all directions for a part of *every* day. Strangely, this is closer to tropical conditions than to neighbouring temperate ones. Although surface radiation *deficits* plunge to 70 W m^{-2} in the heart of the Arctic during January, they climb to respectable *surpluses* of 150 W m^{-2} in July. The advection of heat energy from the Tropics across mid-latitudes polewards, therefore, is only really important during winter when the troposphere shrinks in thickness at high latitudes. This produces a marked temperature inversion over Arctic ice up to 3000 m deep. In summer, the inversion almost breaks down but remains more isothermal as the troposphere expands, i.e. there are hardly any significant changes in temperature with height, and lapse rates are nearly vertical (see Fig. 3.19).

All this adds up to a strange but crucial feature of weather patterns in polar regions, namely, that they are much more active around their *margins* than in the middle. Like an ice cream, the edges are more sensitive to surrounding mass and heat energy exchanges. Remember, too, that radiation losses are greater off warmer surfaces where energy is available (see pp. 42–43). Surface temp-

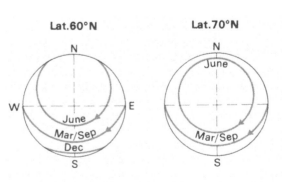
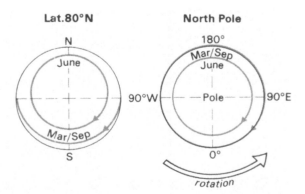

Figure 5.3 Sun paths across skies at high latitudes

eratures are so low that, except for a short time in summer, the heart of either the Arctic or Antarctic has little 'to give'.

Features of temperate weather

It follows that the typical temperate weather patterns already described (see Ch. 4.A) are squeezed between the more active edges of the polar and tropical regions on either side. The former dominates during winter and, fortunately for regions like the British Isles, the less active tropical one exerts more influence in summer.

Dubbed as the temperate zones by the ancient Greeks over 2000 years ago, the northern margins between 50°N and 70°N are simply referred to as 'cold' and the southern zones dipping into subtropical latitudes as being 'warm'. The reverse pattern holds in the southern hemisphere. Northwestern Europe, therefore, falls in the awkward transitional belt where both global and zonal patterns are most delicately poised. To this variety can be added the extra ingredient of being positioned between ocean and continent. Indeed, Nantes in France is the 'centre of gravity' of the world's land mass distribution.

Figure 5.4 shows the broad features arising from Europe's maritime position astride the temperate latitudes. An arrow pointing northwards to Iceland indicates an increase in the overall frequency of mid-latitude depressions passing this way along with the fact that winter routes tend to be more southerly. Broad arrows span the tracks followed by polar and tropical air masses across the Atlantic

(see Figs 3.16 and 4.1). The former approach along the more common cyclonically curved track and the latter drift around the northern edge of the subtropical high-pressure region (STHP) anticyclonically (see Fig. 2.24). This STHP system is more or less permanently stationed in mid-Atlantic over the Azores (see Figs 2.10 and 5.1). Being warmed from below by the Gulf Stream and North Atlantic Drift makes cyclonic polar air unstable, but the tropical air curling into higher latitudes remains more stable (see Ch. 4.C). Their precise tracks are important to weather forecasters. For example, a depression taking polar air far south before curving it back towards Britain allows it to become warmer, moister and highly unstable on its long sea 'voyage'. Such returning polar maritime air produces prolonged heavy rainfalls at all times of year.

The zonal tracks of mid-latitude depressions across Europe are often blocked in winter when the snows arrive on the mountains and the interior cools to set up stable cold-core highs (see Fig. 2.22). These shallow systems squat stubbornly in the positions indicated on Figure 5.4, causing the more ephemeral depressions to divert around them on more northerly or southerly tracks than usual. The Scandinavian high (**A**) and the central European high (**C**) tend to occur in spring and autumn respectively. Lingering winter snows at high latitudes and early snows at high altitude reduce radiation budgets which lead to inversions with subsiding and outblowing air. Similarly, the massive Russian or Siberian high (**B**) takes over in deep winter once the interior plains of Eurasia chill off. These blocking continental highs may occasionally shift or relax during the winter period and break down altogether in summer when the land warms up. The timing of their arrival and departure, as well as the positions they take up, are critical to the onset of spring and autumn in Britain.

Now, imagine what happens when the upper westerlies begin looping as Rossby waves (see Fig. 2.25). A typical sequence known as the **index cycle** is shown in Figure 5.5. These can last up to four to six weeks giving *spells* of consistent weather. Notice how zonal components of wind speeds (west to east) drop as the wave loops and flows become more meridional (north to south). The high zonal index with winds approaching steady jet stream velocities (see Ch. 2.G) can be closely linked with the typical pattern shown on Figure 5.4. As the zonal index decreases, however, grea-

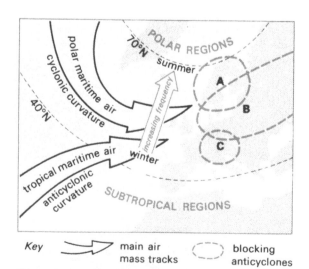

Key: main air mass tracks / blocking anticyclones

Figure 5.4   Air mass tracks to Europe

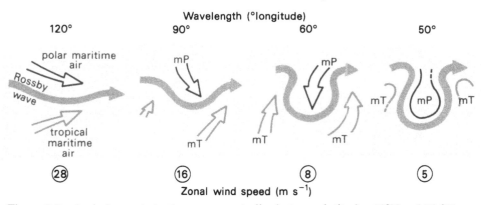

Figure 5.5 An index cycle in the upper westerlies between latitudes 40°N and 70 °N

ter contrasts between alternating northerly and southerly air masses occur. Furthermore, the tight loops are ideal for locking into phase with surface systems as described at the end of Chapter 2. These loops can establish persistent types of weather for several days on end from either northerly or southerly sources. Until they cut themselves off and suddenly return to strong zonal flows with changeable weather, different localities at similar latitudes experience contrasting conditions.

There is growing evidence from actual records and historical information that similar cycling of the circum-polar vortex occurs over longer periods of time causing rhythmic climatic changes. Among the apparent cycles recognised are ones at about 20 years, 80 years, 180 years, 400 years, 2500 years and then those of the last Ice Age during Pleistocene times. Unravelling such cycles within cycles is very difficult, but a pattern is emerging according to Professors Lamb and Bryson. Figure 5.6 illustrates two straightforward possibilities; the high zonal flow produces the sort of weather patterns that have dominated the first half of this century during the two climatic 'normal' periods since 1901 and the low zonal flow is thought to be taking over. Notice how, if this 'locked in' over Europe as shown, the British Isles could experience more persistent northerlies with at least colder and snowier winters. On the other hand, regions in anticyclonic loops would become warmer and drier. Moreover, the pushing further southwards of cold air could be related to the Equatorward shift of the tropical rain-bearing winds mentioned at the end of the previous section. It is also referred to in the following section.

The active margins of the polar ice cap are closely watched as tell-tales for changing patterns in middle and high latitudes. For example, with strong zonal flow, the break-up of pack ice within sheltered seas such as Hudson Bay and Baffin Bay would be earlier in Spring (see Photo. 18). As the zonal index decreases, the break-up might be expected to occur later. Since there are also good naval records of sea ice breaking up in such regions, it is possible to extend the study backwards to get a more complete picture of what is happening to our weather.

In concluding this section yet stimulating further thought, it has been argued that continued 'bumper' harvests in the Middle West of America, as against recent failures of corresponding crops like wheat in Russia, are at least partly a consequence of such climatic changes in temperate regions. The implications of such theories are immensely important!

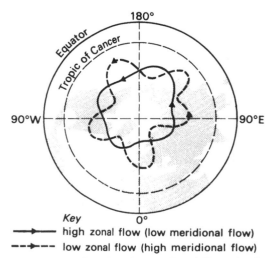

Key
— high zonal flow (low meridional flow)
---- low zonal flow (high meridional flow)

Figure 5.6 Zonal and meridional flows in the upper westerlies

Photograph 18   Sea ice. Arctic ice breaking up in spring in Baffin Bay at latitude 70°N. The bright chunky blocks are bergs of land ice that have calved off the Greenland ice cap and the dull platy patches are floes of sea ice as the floating pack melts. (Seen from an altitude of 10 000 m)

Photograph 19   Mountain ice. Alpine snow and ice in summer in the Swiss Alps at latitude 46°N. The moraines in the foreground below the present day permanent snow line show that glaciers have melted and receded in the past 100 years. The snow line coincides with the cloud base on the opposite side of the valley at an altitude of nearly 3000 m (see Fig. 5.7)

C   Patterns for the future

From space we are now able to survey the entire Earth and its atmosphere by remote-sensing techniques. These offer unprecedented scope for watching most atmospheric processes. Weather satellites also crown Man's historical quest for more information about the air *in depth*.

The ancient Greeks 'coined' the terms meteorology (meaning things above) and climatology (meaning the sloping surface) because their pioneer studies of the atmosphere were confined to observations from the ground. Subsequently, mountains were climbed in search of more knowledge as they are the most sensitive places that can be reached for direct and continuous earthbound observations. By relating measurements from mountain-top observatories and satellites, it is hoped that the weather patterns of the future will become clearer.

The final illustration, Figure 5.7, shows that atmospheric and landscape processes are closely tied together at all space and time scales. Like an elbow dipped into water to sense its temperature, the high mountain that tops the snow line reaches the most diagnostic levels in the air. Layers of glacier ice contain fossil evidence of past processes in immense detail and the advances and retreats of their snouts elevates them to giant natural ther-

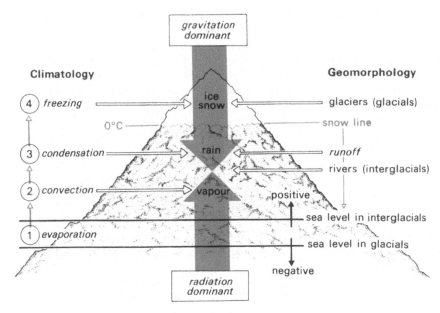

Figure 5.7 The geography of climatic change

mometers responding to shorter-term climatic changes. If or when the next glacial phase comes, we should expect to be warned by the mountain glaciers first (see Photo. 19).

This brief ascent of the mountain would be incomplete without mention of the similarity between the last diagram (Fig. 5.7) and the very first illustration in the book (Fig. 1.1). Both indicate that atmospheric processes and, therefore, climatic changes could arise from either *external* or *internal* causes. In reality, of course, they are more likely to be a combination of the two controls. The former would entail variations in the amounts of solar radiation reaching the atmosphere and the latter would depend upon alterations of the Earth's surface itself. Six of the many possibilities being carefully researched are now reviewed.

External processes
It is apparent that solar radiation will vary by any of the following means:

(a) *Changes in the distribution and fusion of hydrogen ions on the Sun leading to fluctuations in the production and output of solar energy into space.* Such short-term cycles are evident with Sun-spot activity every 11·2 years, but possible long-term changes remain debatable.

(b) *Changes in the Earth's orbit around the Sun to alter the distance between them or the transparency of the intervening space.* Astronomical measurements show that planets do have eccentric orbital paths over long periods of time and that these regularly pass through clouds of cosmic particles which screen radiation.

(c) *Changes in the tilt of the Earth's axis of rotation relative to the plane of its orbit around the Sun.* For instance, if the present angle of tilt of $66\frac{1}{2}°$ alters, then seasonal balances of radiation at the surface would change. It is known that the axis of rotation 'wobbles' slightly. Furthermore, some geologists suppose that the problematic Precambrian glaciation in the southern hemisphere can be explained by aligning the axis of that time along that of the Equator today, i.e. a 90° shift in the Earth's obliquity.

Internal processes
Apart from the possible variations in Earth's gravitation that have been suggested by some geophysicists, less speculative reasons for climatic change are as follows.

(a) *Changes in the composition of the air brought about by geological and biological agencies.*

For example, volcanic activity is linked with orogenic or mountain-building cycles to vary amounts of CO_2, water vapour and dust exhaled into the atmosphere. The evolution of plants and animals has a demonstrable connection with the formation of the atmosphere and the proportions of radiation absorbing gases present in the air.

(b) *Changes in the geographical distribution of land and sea, particularly the elevation and alignment of high mountain ranges.* Continental drift owing to plate tectonics is now a well established explanation for ancient climates (palaeoclimates) during geological time. There is reason to suppose that different atmospheric and oceanic circulations existed in the past, particularly from wind-blown fossil evidence such as pollen grains and sands preserved in undisturbed sediments on the sea floors.

(c) *Changes brought about by Man altering the surface features of the ground and by dumping wastes into the air which are not recycled.* As explained already, there is growing evidence to show that many human activities are causing the atmosphere to get warmer.

It is clear that we know a little about many of the above causes of climatic change but virtually nothing about how they might be related in altering the global patterns of atmospheric processes. Most geographers look for evidence of change from landscapes whereas meteorologists and climatologists search for empirical explanations from the atmosphere itself, notably in the behaviour of the general circulation. Some have enlikened the Hadley cells to giant fly wheels in the atmospheric machine, with the mid-latitude waves as the gearing system (see Fig. 1.10). When the atmosphere warms up, the former turn faster to generate more energy, and drive the mid-latitude westerlies harder. As illustrated earlier (see p. 86) this would entail them having a high zonal index as evident throughout much of this century. Continued warming, especially in the tropics, could lead to overheating which would be countered by the mid-latitude waves looping more in order to facilitate the exchange of colder polar and warmer tropical air meridionally. This, in turn, would gear down the system to brake the Hadley cells. More precipitation and then cooling would take over. Such cycles may explain the succession of warm and rainy phases (interglacials) and cold but, probably, drier phases (glacials) during the past 2·5 million years of geological time. The balance of this evidence points to a return to glacial conditions in future.

Paradoxically, then, some researchers have argued that the present warming could be the prelude to greater precipitation, faster cooling and the more rapid onset of the next glaciation. It is reckoned that a global temperature decrease of less than 5°C would see a return of the glaciers to most of Europe and North America. This, after all, is what really happens when a room overheats and the windows are opened to chill the air. Likewise, it would be remarkable if the atmospheric machine described throughout this book did not experience fluctuations in heating and cooling of its own making.

By unravelling the fundamental mechanisms responsible for atmospheric processes, geographers are better able to reach more profound explanations of Man's relationships with his environment, whether past, present or future. Neither is there a sounder reason for their study than the sobering thought that many aspects of climates are still poorly understood or appreciated. Despite the advanced state of modern technology, we have no means of knowing for sure whether we are free from the wonders and whims of the weather.

Geographers who choose to ignore a basic study of atmospheric processes, therefore, must conclude that their synthesis of any environmental information is incomplete.

Further Reading

A Books covering a wide range of examples

Boucher, K. 1975. *Global climate*. London: English Universities Press.

Geiger, R. 1965. *The climate near the ground*. Cambridge, Mass.: Harvard University Press.

Lockwood, J. G. 1974. *World climatology*. London: Edward Arnold.

Riley, D. and L. Spolton. 1974. *World weather and climate*. Cambridge: Cambridge University Press.

Taylor, J. A. and R. A. Yates. 1967. *British weather in maps*. London: Macmillan.

B Books of further academic interest

Theory

Barry, R. G. and R. J. Chorley. 1971. *Atmosphere, weather and climate*. London: Methuen.

Barry, R. G. and A. H. Perry. 1973. *Synoptic climatology*. London: Methuen.

Chandler, T. J. 1972. *Modern meteorology and climatology*. London: Nelson.

Corby, G. A. (ed.) 1970. *The global circulation of the atmosphere*. London: Royal Meteorological Society.

Crowe, P. R. 1971. *Concepts in climatology*. London: Longmans.

Flohn, H. 1969. *Climate and weather*. London: Weidenfeld & Nicholson.

Goody, R. M. and J. C. G. Walker. 1972. *Atmospheres*. Englewood Cliffs, N.J.: Prentice-Hall.

Lamb, H. H. 1972. *Climate: past, present and future*, Vol. 1. London: Methuen.

Lockwood, J. G. 1979. *Causes of climate*. London: Edward Arnold.

Stringer, E. T. 1972. *Foundations of climatology*. San Francisco: W. H. Freeman.

Measurement techniques

Barrett, E. C. 1974. *Climatology from satellites*. London: Methuen.

Bruce, J. P. and R. H. Clark. 1966. *Introduction to hydrometeorology*. Oxford: Pergamon.

Hanwell, J. D. and M. D. Newson. 1973. *Techniques in physical geography*. London: Macmillan.

Stringer, E. T. 1972. *Techniques in climatology*. San Francisco: W. H. Freeman.

Applied climatology

Atkinson, B. W. 1968. *The weather business*. London: Aldus.

Bradshaw, M. 1977. *Earth, the living planet*. London: Hodder & Stoughton.

Goudie, A. S. 1977. *Environmental change*. Oxford: Clarendon.

Lacy, R. E. 1977. *Climate and building in Britain*. London: HMSO.

Mason, B. J. 1962. *Clouds, rain and rainmaking*. Cambridge: Cambridge University Press.

Maunder, W. J. 1970. *The value of the weather*. London: Methuen.

Wickham, P. G. 1970. *The practice of weather forecasting*. London: HMSO.

Index

References to illustrations, figures and photographs, are given in *italics*

Milton Keynes UK
Ingram Content Group UK Ltd.
UKHW051855071024
449327UK00025B/1972